CELESTIAL
NAVIGATION

CELESTIAL NAVIGATION

Captain Joe Thompson's
Cookbook Method

Captain
Joseph E. Thompson

David McKay Company, Inc.
New York

Library of Congress Cataloging in Publication Data
Thompson, Joseph E
Celestial navigation
1. Navigation.   2. Nautical astronomy.   I. Title.
VK555.T46   623.89   80-26946
ISBN 0-679-50965-8

2 3 4 5 6 7 8 9 10
MANUFACTURED IN THE UNITED STATES OF AMERICA

. . . to my Pop whose passing caused me to seek escape from the daily routine, the result of which is this book. And to my friends, for being my friends.

JET

CONTENTS

CELESTIAL
NAVIGATION

1.

INTRODUCTION

Introduction

This "cookbook" of celestial navigation was born in frustration. Back when I was first studying the subject, everyone assured me that there was no great need to *understand* what I was doing. It was necessary only to learn the simple mechanics of sextant use and sight reduction in order to obtain lines of position or a fix. All assurances were that with a little study I could "cookbook it"; just follow the steps and my answer would be right before me.

Nevertheless, all the books I could borrow or buy, and a considerable amount of personal instruction, seemed too complex for simple understanding. For someone who wanted merely to learn how to navigate, there was too much emphasis on: (1) the mathematical formulas of spherical trigonometry, (2) the effect of crossing the international date line from east to west at midnight, (3) the technical differences between various historic reduction tables and methods of computation, and (4) the glories of the latest pocket calculator with its memory functions.

Nowhere could I find a simple description of the steps required to locate one's position without having to wade through an ocean of technical and jargon-filled explanations of celestial navigation. Even the most intelligible texts first had to stop the world and then have the heavenly bodies revolve around a stationary earth.

Perseverance bordering on masochism finally prevailed. Eventually, as the vocabulary became more familiar, I was able to separate the "need to know" from the "nice to know" and the subject became more manageable. Sometime afterward I reviewed the material I had studied earlier. This included some 2,000 pages of *American Practical Navigator,* or "Bowditch," as it is commonly called. Surprise and satisfaction! Much of it made sense. Finally, even the so-called basic books seemed simple and, indeed, basic.

This book is to be the complete and simple step-by-step "how to" book that I sought. It is as straightforward as possible in each mechanical step of all the major celestial navigation requirements. As far as possible, all theory has been eliminated or reduced to a minimum. A brief theoretical discussion is provided toward the end of the book for those who want to learn *why* after learning *how*.

As to its structure, this book is to be used in conjunction with *The Nautical Almanac* and the H.O. Pub. No. 249 series, Volumes I, II, and III, *Sight Reduction Tables for Air Navigation*. (See photograph.) It is specifically intended that the reader should become familiar with the planet chart and text in the front of *The Nautical Almanac* and the "Explanation" section and the star charts at the back. These few pages are particularly well done and have permitted this book to avoid needless duplication of text and examples. Similarly, there are several pages of clearly written instructions and information in the opening pages of the *Sight Reduction Tables*.

In addition, the use of *The Nautical Almanac* as a companion to this book permits reference to a current document. Thus it is not necessary to duplicate pages from the almanac for purposes of illustration. Such an arrangement, in effect, creates space in this text for step-by-step explanation of the mechanics of reduction. While these procedures are often fairly similar for the various celestial bodies, the differences

are often critical, subtle, or technical (read "tricky") enough to justify substantial repetition and detail.

Volume I of *Sight Reduction Tables* is devoted to selected stars and is published at five-year intervals. Volume II, for latitudes 0°-39°, and Volume III, for latitudes 40°-89°, provide complete data for the reduction of sights of sun, moon, planets, and stars within the declination range of these bodies, extending from the zenith to the horizon below.

To emphasize, then, since it is absolutely necessary for either the experienced navigator or the student to have *The Nautical Almanac* and the *Sight Reduction Tables* at hand when performing celestial navigation, this book is designed with the assumption that these references are immediately available. This book is intended to be as valuable in actual use as in study. As with other cookbooks of the standard type, one is expected to have the ingredients, not just the instructions, at hand when preparing the meal.

Other types of almanacs and reduction tables are available for those who might wish them. It is my conclusion (and one shared by many others) that the combination of *The Nautical Almanac* and the H.O. 249 series is unbeatable. These two references offer superior compactness, ease, and speed of handling, use, cost, and functional accuracy. Therefore, this book is based on those publications.

As indicated, the purpose and scope of this book are limited. This is not a treatise on why things are as they are, or even why celestial navigation functions and helps mariners in small boats cross large oceans with confidence and accuracy. Such material is very interesting and worthwhile, but it can be studied and understood separately from the mechanics of sextant operation, the use of sight reduction tables, and the determination of one's location.

It is also fair to say that this book presumes certain understanding on the part of the reader. The ability to read a compass, understand standard navigational charts, and plot a true course are required even for coastal piloting. Thus it is anticipated that those skills have been mastered before one undertakes the study of celestial navigation.

Finally, I would like to share my hopes as to what this book may accomplish. First, I hope that it simplifies what the beginning student of celestial navigation is trying to do . . . to learn as easily as possible how to locate his position with the use of a sextant. Second, I hope that more advanced navigators and those who have let their skills get rusty through months or years of idleness will find it easier to develop or recapture the skills required to accomplish the long cruise that is fermenting in their souls. And third, I hope and believe that this description of the mechanics of celestial navigation will be of considerable assistance to the old salt who has long known and understood such navigation but who wishes to brush up on the mechanics or who, through fatigue or for another reason, is having trouble making his wearied efforts produce the answer to the only question in his mind, "Where am I?"

Good luck.

2.

THE SEXTANT: COMMENTS ON SELECTION AND USE

The Sextant: Comments on Selection and Use

Drawing from general experience in life, I have observed that comments on another man's woman and why he may have selected her are fraught with misunderstanding if not outright danger. So it is appropriate that sextant begins with "sex" and that I exercise the judgment of my years. I shall be entirely evasive with regard to brand names and focus on common characteristics that may be considered in sextant selection.

As in the case of everything else, sextants vary widely in cost, style, quality, materials, reliability, serviceability, and extra-cost add-ons. Purchase decisions are generally a combination of intended use, ego, affordability, style, and judgment—good and bad.

Some people want everything—the finest quality, a bubble to serve as an artificial horizon, a seven-power scope to see the smallest star, a night light to illuminate the scale, and a lovely mahogany case in which to store and carry their most valued possession. Others, perhaps most mindful of cost, will accept the cheapest thing they can buy as long as the advertisement assures them that it has been used to cross both oceans.

In theory you get what you pay for. But consider a comparison with automobiles. As a means of transportation, is a $60,000 auto really worth 10 times more than one costing $6,000? Not for me. On a smaller scale, though perhaps to a similar degree, one may compare sextants. Let us examine the characteristics and my opinions, and then you may choose.

A. **Optics.** The big cost differential among the better sextants generally is related to the quality of optical lenses and whether several powers or scopes are available. High-power scopes (4X and more) may be of use under ideal conditions or on a supertanker, which is stable. On small boats they are very difficult to use if there is substantial movement. A good scope of about 2.5 to 4.0 power is a considerable asset if you ever plan to do much besides noon sun shots. For my primary sextant, on a scale of 1 to 10, I give high-quality optics with a lens of modest power a score of 8.

B. Night Light.   While having a light built into the sextant is both sexy and useful under certain conditions, there is always the danger of forgetting to remove the batteries when the instrument is not in use for extended periods and corroding the whole thing. One sextant has a light as a clip-on feature; it could be worthwhile. For the added cost, however, I can use a penlight. I give this a 4 on my scale.

C. Carrying Case.   This is an ego item for me. For the primary sextant, I'll admit to giving it a score of 7. Perhaps you can have pure function prevail. In that case, all cases work pretty well, but select one with a good latch that will not fall open accidentally.

D. Bubble Horizon.   I have so little experience with sextants with built-in artificial horizons, or "bubbles," that I am not competent to pass judgment. Friends advise that they are nice until dropped or knocked out of adjustment. Old military bubble sextants are notoriously difficult and expensive to have repaired. On a scale of 1 to 10, I give this a 2 and forget it.

E. Material, Size, and Quality.   In regard to material, especially traditional brass construction, some of the finest sextants have a failing in that they are *much* too heavy. Some of the lightweight alloy ones tend to corrode. While I have never had any difficulty with moderate-to high-quality instruments, some Japanese models are reported to have manufacturing deficiencies.

As to size, the so-called three-quarter-size sextants are comparatively lighter than the larger ones. The resulting decrease in the size of the scale, however, makes their measurements slightly more difficult to judge.

My conclusions on sextant quality are mixed, but my recommendation is to lean toward paying the price for a good one and hope that the difference is justified through years of reliable service. For a primary sextant, of a possible score of 10, I give quality a 7.

F. Serviceability.   The mystique that surrounds the art of sextant adjustment makes the use of the instrument look like child's play. No doubt the original head of the guild was a close friend of Blackbeard, if one judges by the prices charged. On the other hand, high precision and quality control command fair compensation. In general, when it comes to doing work yourself on a particularly fine instrument built to exacting tolerances and finely adjusted with a maze of carefully tightened screws, the best advice is "don't." Without instructions or a good understanding of what you are doing, your chances of breaking something, particularly a mirror, are excellent. On the other hand, the American-made plastic sextants, while perhaps more prone to developing errors, are cleverly designed and easily adjusted in seconds with a thumbscrew.

G. Sextant Error. All sextants may be expected to have errors. Four types—prismatic, shade, graduation, and centering—are generally considered nonadjustable and are collectively referred to as "instrument error." Instruments distributed by better manufacturers customarily attach a table inside the sextant case indicating the error for various readings of the arc. To make the instrument correction, therefore, you reverse the sign of the error. With most sextants, instrument error is quite small.

While the following chapters do not make further specific reference to instrument error correction, it may be included in each calculation at the point of entering the index error adjustment.

Certain adjustable errors in the sextant are related to perpendicularity and parallelism of the frame, mirrors, and horizon glass. Correction of these errors is accomplished by the adjustments mentioned under "Serviceability" above.

A most common error among sextants is called "index error." This error is determined with each use of the sextant and can be compared to the tuning of a musical instrument. The index error is observed by setting the index arm at zero degrees and aiming the instrument at the horizon. The horizon will probably appear slightly offset between the mirror surface and the clear glass. See Figure 2.1. To measure this error, line up the real and mirror horizons so that both horizons form one straight line. Now read the sextant scale. If it reads other than zero, the value shown is index error. Index error is applied to the sextant reading by reversing the sign as an Index

FIGURE II.1

SEXTANT INDEX ERROR

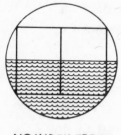 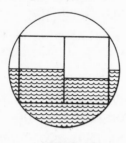 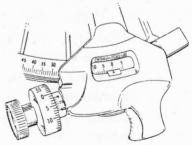

NO INDEX ERROR

ERROR OFF
THE ARC

ERROR ON
THE ARC

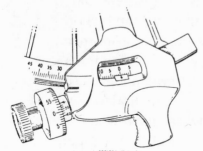

(b) Reading is 0°3'20" off the Arc

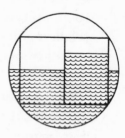

(a) Reading is 0°2'10" on the Arc

Correction (IC). If the error is *on* the arc (that is, toward the long part of the scale), subtract; if it is *off* the arc (that is, on the short part of the scale), add. In the subsequent text, IC is included as the first adjustment to the sextant reading.

Another error, that of collimation, results if the telescope is not parallel to the instrument. Better sextants have screws located on the collar holding the telescope, which may be adjusted to correct this problem.

H. **Atmospheric Errors.** Large errors may appear in two common situations. While these errors are not the fault of the sextant, this would appear a good opportunity to discuss them.

In the first situation, the body being sighted is generally only 15° or less above the horizon. Although the use of low-altitude tables is explained elsewhere (Chapter 9, "Navigation without Sextant"), it is worthwhile to remember that the effects of unusual atmospheric conditions, passing squall lines and such, can produce abnormally large errors. The adverse effects of such conditions are generally most severe at lower altitudes. Thus, most navigators prefer not to shoot bodies below 15° of altitude. When they do, they view the results with caution.

The second situation is encountered when using bodies with altitudes in excess of 60°— when the vessel is in the lower latitudes, for example, and the navigator is attempting to shoot a noon sight with the sun almost directly overhead. In an effort to understand and overcome this phenomenon, I have seen experienced navigators shoot high-altitude noon sights from a predetermined position and consistently get unsatisfactory results. Even under such controlled conditions, the errors often approximate 20 miles, although a geometric technique using the radius (or zenith distance) from multiple high altitude sights taken right before, during, and after the local noon can be used to draw small enough circles to produce a fix. Anyway, of the 90° of arc from the horizon to your zenith, the 45° from 15° to 60° may be considered as most desirable when determining the bodies or the times for your sightings.

I. **Other Sextant Applications.** The use of the sextant as a device for pilotage is perhaps even more common than its use for celestial navigation. Although this application is beyond the scope of this book, the reader is urged to develop familiarity with the sextant as a device for measuring distance off from an object whose height is known (distance by vertical angle) and in determining his position near shore by holding the instrument horizontally and measuring angles between terrestrial objects (position by three-point fix). These are simple applications of the sextant. Anyone seeking to develop celestial navigation skills should certainly take an hour to study and learn these techniques. Often, especially near sunrise or sunset, and occasionally with the moon, by taking a shot over the ship's compass and recording the time and compass bearing of the body, the deviation of the compass can be determined within several degrees. As the correct zenith of the body will be produced when reducing the sight, any difference must be due to variation and deviation. Since the chart will show you the variation for the area, any further difference is the deviation of the compass on that bearing.

J. **Sextant Use.**   Although the use of the sextant, insofar as actual handling is concerned, is doubtless passed on from friend to friend rather than through books, there are several basic points that must be made.

When you find the celestial body in the eyepiece, move the index arm on the sextant until the body being observed in the mirrored surface just barely touches the horizon. It is more common to sink the body, or the lower portion (limb) of the sun or moon, into the water than it is to keep it too high. Also, in general it is best for the observer to stand in a position as high as conveniently practical in order to have his head as high as possible. This causes the horizon to be farther away and increases dip, thus giving a more certain image. If there are pronounced waves or swells, be certain to take the shot at the top of the wave or swell rather than in the trough. Under severe conditions, it can be very difficult to get a sextant shot in which you have confidence.

To make certain that the object is just touching the horizon, it is necessary to rock, or rotate, the sextant from left to right around the axis of the telescope. See Figure 2.2. The purpose is to ensure that the image is directly beneath the object rather than slightly to one side, which would have the effect of increasing the angle reading.

It is also possible to hold the sextant upside down and adjust it to raise the horizon up to the body. This is seldom done except with

FIGURE II.2

ROCKING THE SEXTANT

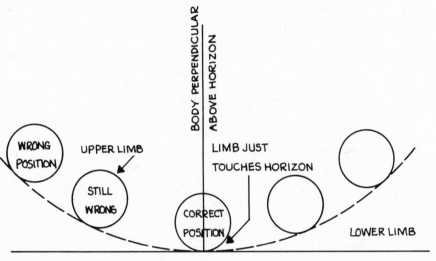

ROCKING SEXTANT FOR FULL MOON OR SUN, USING LOWER LIMB

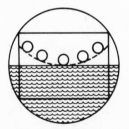

ROTATE SEXTANT AROUND AXIS
OF TELESCOPE

stars, which tend to get lost by slipping out of view on the telescope during heavy weather.

When practicing with the sextant, be sure to try various combinations of filters, or shades, to learn which ones you prefer under different circumstances. Certain combinations are particularly effective, not only for bright sunlight, but also for clouds, mists, and some fog conditions.

Something many people fail to appreciate is that you do not need an uninterrupted horizon to use your sextant. Bowditch contains a table entitled "Dip of the Sea Short of the Horizon." This table indicates, for various heights of eye of the navigator, an adjusted figure for dip that must be used if the horizon is obstructed within stated distances. The interesting thing, though, is that with a height of eye of 10 feet, for example, if the obstruction, or shoreline, or whatever, is 3.7 nautical miles or more distant, the obstruction can be ignored. For that height of eye the observer's horizon is closer than the object. This applies even if land is clearly visible on the other side of the bay. Using this information and the table can greatly increase the opportunity for practice. See Figure 2.3.

Speaking of practice, a generous amount of it is required in order to become competent

FIGURE II-3

TABLE 22

Dip of the Sea Short of the Horizon

Dis-tance	Height of eye above the sea, in feet										Dis-tance
	5	10	15	20	25	30	35	40	45	50	
Miles	′	′	′	′	′	′	′	′	′	′	*Miles*
0. 1	28. 3	56. 6	84. 9	113. 2	141. 5	169. 8	198. 0	226. 3	254. 6	282. 9	0. 1
0. 2	14. 2	28. 4	42. 5	56. 7	70. 8	84. 9	99. 1	113. 2	127. 4	141. 5	0. 2
0. 3	9. 6	19. 0	28. 4	37. 8	47. 3	56. 7	66. 1	75. 6	85. 0	94. 4	0. 3
0. 4	7. 2	14. 3	21. 4	28. 5	35. 5	42. 6	49. 7	56. 7	63. 8	70. 9	0. 4
0. 5	5. 9	11. 5	17. 2	22. 8	28. 5	34. 2	39. 8	45. 5	51. 1	56. 8	0. 5
0. 6	5. 0	9. 7	14. 4	19. 1	23. 8	28. 5	33. 3	38. 0	42. 7	47. 4	0. 6
0. 7	4. 3	8. 4	12. 4	16. 5	20. 5	24. 5	28. 6	32. 6	36. 7	40. 7	0. 7
0. 8	3. 9	7. 4	10. 9	14. 5	18. 0	21. 5	25. 1	28. 6	32. 2	35. 7	0. 8
0. 9	3. 5	6. 7	9. 8	12. 9	16. 1	19. 2	22. 4	25. 5	28. 7	31. 8	0. 9
1. 0	3. 2	6. 1	8. 9	11. 7	14. 6	17. 4	20. 2	23. 0	25. 9	28. 7	1. 0
1. 1	3. 0	5. 6	8. 2	10. 7	13. 3	15. 9	18. 5	21. 0	23. 6	26. 2	1. 1
1. 2	2. 9	5. 2	7. 6	9. 9	12. 3	14. 6	17. 0	19. 4	21. 7	24. 1	1. 2
1. 3	2. 7	4. 9	7. 1	9. 2	11. 4	13. 6	15. 8	17. 9	20. 1	22. 3	1. 3
1. 4	2. 6	4. 6	6. 6	8. 7	10. 7	12. 7	14. 7	16. 7	18. 8	20. 8	1. 4
1. 5	2. 5	4. 4	6. 3	8. 2	10. 0	11. 9	13. 8	15. 7	17. 6	19. 5	1. 5
1. 6	2. 4	4. 2	6. 0	7. 7	9. 5	11. 3	13. 0	14. 8	16. 6	18. 3	1. 6
1. 7	2. 4	4. 0	5. 7	7. 4	9. 0	10. 7	12. 4	14. 0	15. 7	17. 3	1. 7
1. 8	2. 3	3. 9	5. 5	7. 0	8. 6	10. 2	11. 7	13. 3	14. 9	16. 5	1. 8
1. 9	2. 3	3. 8	5. 3	6. 7	8. 2	9. 7	11. 2	12. 7	14. 2	15. 7	1. 9
2. 0	2. 2	3. 7	5. 1	6. 5	7. 9	9. 3	10. 7	12. 1	13. 6	15. 0	2. 0
2. 1	2. 2	3. 6	4. 9	6. 3	7. 6	9. 0	10. 3	11. 6	13. 0	14. 3	2. 1
2. 2	2. 2	3. 5	4. 8	6. 1	7. 3	8. 6	9. 9	11. 2	12. 5	13. 8	2. 2
2. 3	2. 2	3. 4	4. 6	5. 9	7. 1	8. 3	9. 6	10. 8	12. 0	13. 3	2. 3
2. 4	2. 2	3. 4	4. 5	5. 7	6. 9	8. 1	9. 2	10. 4	11. 6	12. 8	2. 4
2. 5	2. 2	3. 3	4. 4	5. 6	6. 7	7. 8	9. 0	10. 1	11. 2	12. 4	2. 5
2. 6	2. 2	3. 3	4. 3	5. 4	6. 5	7. 6	8. 7	9. 8	10. 9	12. 0	2. 6
2. 7	2. 2	3. 2	4. 3	5. 3	6. 4	7. 4	8. 4	9. 5	10. 6	11. 6	2. 7
2. 8	2. 2	3. 2	4. 2	5. 2	6. 2	7. 2	8. 2	9. 2	10. 3	11. 3	2. 8
2. 9	2. 2	3. 2	4. 1	5. 1	6. 1	7. 1	8. 0	9. 0	10. 0	11. 0	2. 9
3. 0	2. 2	3. 1	4. 1	5. 0	6. 0	6. 9	7. 8	8. 8	9. 7	10. 7	3. 0
3. 1	2. 2	3. 1	4. 0	4. 9	5. 9	6. 8	7. 7	8. 6	9. 5	10. 4	3. 1
3. 2	2. 2	3. 1	4. 0	4. 9	5. 7	6. 6	7. 5	8. 4	9. 3	10. 2	3. 2
3. 3	2. 2	3. 1	3. 9	4. 8	5. 7	6. 5	7. 4	8. 2	9. 1	9. 9	3. 3
3. 4	2. 2	3. 1	3. 9	4. 7	5. 6	6. 4	7. 2	8. 1	8. 9	9. 7	3. 4
3. 5	2. 2	3. 1	3. 9	4. 7	5. 5	6. 3	7. 1	7. 9	8. 7	9. 5	3. 5
3. 6	2. 2	3. 1	3. 8	4. 6	5. 4	6. 2	7. 0	7. 8	8. 6	9. 4	3. 6
3. 7	2. 2	3. 1	3. 8	4. 6	5. 4	6. 1	6. 9	7. 7	8. 4	9. 2	3. 7
3. 8	2. 2	3. 1	3. 8	4. 6	5. 3	6. 0	6. 8	7. 5	8. 3	9. 0	3. 8
3. 9	2. 2	3. 1	3. 8	4. 5	5. 2	6. 0	6. 7	7. 4	8. 1	8. 9	3. 9
4. 0	2. 2	3. 1	3. 8	4. 5	5. 2	5. 9	6. 6	7. 3	8. 0	8. 7	4. 0
4. 1	2. 2	3. 1	3. 8	4. 5	5. 1	5. 8	6. 5	7. 2	7. 9	8. 6	4. 1
4. 2	2. 2	3. 1	3. 8	4. 4	5. 1	5. 8	6. 5	7. 1	7. 8	8. 5	4. 2
4. 3	2. 2	3. 1	3. 8	4. 4	5. 1	5. 7	6. 4	7. 0	7. 7	8. 4	4. 3
4. 4	2. 2	3. 1	3. 8	4. 4	5. 0	5. 7	6. 3	7. 0	7. 6	8. 3	4. 4
4. 5	2. 2	3. 1	3. 8	4. 4	5. 0	5. 6	6. 3	6. 9	7. 5	8. 2	4. 5
4. 6	2. 2	3. 1	3. 8	4. 4	5. 0	5. 6	6. 2	6. 8	7. 4	8. 1	4. 6
4. 7	2. 2	3. 1	3. 8	4. 4	5. 0	5. 6	6. 2	6. 8	7. 4	8. 0	4. 7
4. 8	2. 2	3. 1	3. 8	4. 4	4. 9	5. 5	6. 1	6. 7	7. 3	7. 9	4. 8
4. 9	2. 2	3. 1	3. 8	4. 3	4. 9	5. 5	6. 1	6. 7	7. 2	7. 8	4. 9
5. 0	2. 2	3. 1	3. 8	4. 3	4. 9	5. 5	6. 0	6. 6	7. 2	7. 7	5. 0
5. 5	2. 2	3. 1	3. 8	4. 3	4. 9	5. 4	5. 9	6. 4	6. 9	7. 4	5. 5
6. 0	2. 2	3. 1	3. 8	4. 3	4. 9	5. 3	5. 8	6. 3	6. 7	7. 2	6. 0
6. 5	2. 2	3. 1	3. 8	4. 3	4. 9	5. 3	5. 7	6. 2	6. 6	7. 1	6. 5
7. 0	2. 2	3. 1	3. 8	4. 3	4. 9	5. 3	5. 7	6. 1	6. 6	6. 9	7. 0
7. 5	2. 2	3. 1	3. 8	4. 3	4. 9	5. 3	5. 7	6. 1	6. 5	6. 9	7. 5
8. 0	2. 2	3. 1	3. 8	4. 3	4. 9	5. 3	5. 7	6. 1	6. 5	6. 9	8. 0
8. 5	2. 2	3. 1	3. 8	4. 3	4. 9	5. 3	5. 7	6. 1	6. 5	6. 9	8. 5
9. 0	2. 2	3. 1	3. 8	4. 3	4. 9	5. 3	5. 7	6. 1	6. 5	6. 9	9. 0
9. 5	2. 2	3. 1	3. 8	4. 3	4. 9	5. 3	5. 7	6. 1	6. 5	6. 9	9. 5
10. 0	2. 2	3. 1	3. 8	4. 3	4. 9	5. 3	5. 7	6. 1	6. 5	6. 9	10. 0

in the use of the sextant. To the student I offer the suggestion that it will take approximately 500 recorded observations of the various heavenly bodies to gain facility. You should complete something on the order of 100 sight reductions and lines of position (LOPs) before publicly displaying much confidence, even for favorable conditions. Nevertheless, you can develop this facility in a reasonably short period of time with concentrated practice.

Under difficult circumstances, and when fatigue begins to set in, years of regular experience can make a considerable difference in sextant use and accuracy. Experience is particularly helpful in avoiding mechanical and mathematical errors in completing the paper and chart work. An old adage states that when things get tough it is always better to bet on age and experience than on youth and enthusiasm. This is especially so at sea, particularly with navigation.

Finally, when nearing completion of the voyage and closing with land, the experienced skipper will time his arrival to sight the coast or reach shallow water during daylight. If necessary, he will even spend the night tacking back and forth well offshore to wait for light and a safe finish.

K. Two Sextants.   One of the greatest aids to navigators is the development of the reasonably accurate, low-cost plastic sextant the virtues of the better of these are easy to extol. It is always possible to lose overboard or break your good sextant, and cost often prohibits having in reserve another of equal quality. Often, too, there may be a guest aboard who is curious, or even seriously interested in learning, and to whom it would be nice to be able to offer a sextant. Try as I might, I do not have the courage to offer my good sextant (or binoculars for that matter) to anyone. The plastic ones work practically the same and save me gray hairs.

Finally, I think one of the best justifications for buying a plastic sextant is that you can use it when conditions are bad. Most plastic sextants are quite adequate for offshore service, even if they are not as precisely accurate as the more expensive brass or alloy ones.

Let us be as honest as we are accurate, however. The normal differences in readings of a properly adjusted sextant among experienced navigators is generally much greater than the inherent differences of variously priced instruments themselves. In brief, for practical purposes of locating one's position in a small boat wallowing in sloppy seas, you can forget the manufacturer's claims of greater accuracy for the more expensive unit. The "improvement" in the "accuracy" of your LOP in these conditions is irrelevant. Nonetheless, I confess to using both units from time to time at known locations in order to check the accuracy of each, and were a discrepancy to appear, I would probably select the reading of the "better," or more expensive, of the two as the standard. Also, during favorable conditions when closing with the coast or negotiating dangerous waters, it is normal to reach for the "good" sextant and keep in practice with it.

Under heavy weather conditions, or when the vessel is rolling and pitching and the spray blowing, you can probably obtain readings with the plastic sextant that are just about as accurate as those taken with the others. Due to the weight factor, in some cases the plastic model may be superior. In any event, under such circumstances no one wants to get his "best friend" thoroughly doused. Up steps the plastic backup sextant; it does its work very satisfactorily, getting soaked in the process. Then you take it below, wash it off with fresh water, dry it, and put it back in its plastic carrying case. A good instrument, giving good service.

The Sextant

A final word on sextant use, if the instrument you are using is not equipped with a lanyard, put one on it. The reason for this is so obvious as not to require explanation. Having been warned, if you (or any of your crew) drop the sextant overboard because it has no lanyard, there is no one to blame but yourself.

3.

NOON SHOT
LOCATION OF
POSITION

Noon Shot Location of Position

Items needed: Sextant, *The Nautical Almanac* (N.A.), accurate time, graph paper, scratch paper, and pencil.

A. Note body (sun), date, and deduced latitude and longitude (your dead reckoning or DR position). Select the upper or lower portion (limb) of the sun to bring down to the horizon. Determine height of eye above water.

B. Depending on skill, experience, and conditions, begin taking raw (uncorrected) sextant shots of the sun 10 to 30 minutes before the sun is expected to pass your local meridian.

Record exact watch times and sextant readings until the sun reaches and passes its zenith by 10 to 30 minutes, or until you are sufficiently experienced to get satisfactory results with less time. If taking sextant shots and recording times alone, you may need 20 or more readings.

C. To select the desired sextant reading and time, proceed as follows: On graph paper, place on the left side a vertical scale in degrees and minutes of arc for sextant readings. Across the bottom, place a horizontal scale for time. Arrange, or plot, sextant readings as recorded for degrees and time. In freehand, draw a line connecting the apparent *flow* of the plotted points as in Figure 3.1. On unadjusted data it appears in Figure 3.1 that local noon was at 12 hours, 28 minutes, 54 seconds, and that the raw sextant reading was 54°38.0′. Since the sun may appear to hold a fairly constant altitude for as long as several minutes as it passes your meridian, exact determination of its zenith (highest angle from the horizon) may be difficult. The difficulty may be worse with cloud cover or during bad weather. As it "rises" and "falls" equally before and after it passes your meridian, you may calculate the time by comparing equal points along the

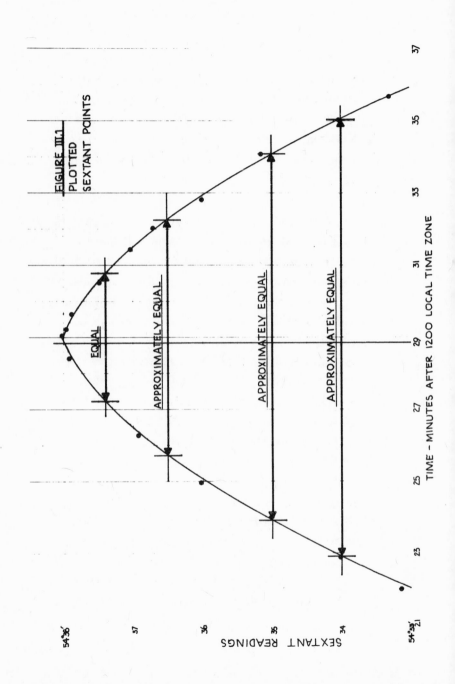

FIGURE III.1
PLOTTED SEXTANT POINTS

TIME – MINUTES AFTER 1200 LOCAL TIME ZONE

SEXTANT READINGS

EQUAL

APPROXIMATELY EQUAL

APPROXIMATELY EQUAL

APPROXIMATELY EQUAL

curves. Thus, noon may actually become more of a construction than an exact sextant reading, although most navigators select one of the sightings closest to the line and work from it, particularly if seeking a latitude line of position only.

D. The time of apparent local noon serves as the basis for determining longitude. First it is necessary to calculate Greenwich mean time (GMT) from the time of the selected sextant shot as follows:

1. Take raw watch time (WT) in hours, minutes, and seconds.

2. Correct for watch error (WE): subtract for fast, add for slow.

3. The sum of WT and WE = ZT (zone time).

4. Add zone description (ZD) to ZT to get GMT. For example, Miami, Florida, is in Zone 5, as zones are 15° wide from Greenwich (80° divided by 15° = 5+ hours); however, during summer, if daylight saving time

(DST) is in effect, add one less hour (i.e., 5 − 1 = 4 hours) to ZT to get GMT.

E. Calculate the Greenwich hour angle (GHA) of the sun as follows:

1. Turn to the proper Greenwich date in the white pages of the N.A. Under the column for the sun, locate the GHA for the whole hour of GMT. Write down the degrees, minutes, and tenths of minutes shown.

2. To interpolate the GHA for the remaining minutes and seconds of GMT, turn to the yellow pages entitled "Increments and Corrections" in the back of the N.A. These pages are marked in minutes of time with seconds of time running down the left-hand columns. Two minutes, with 60 seconds each, are displayed on a page. Having selected the correct minute and second of time, take from the "Sun and Planets" column the degrees, minutes, and tenths of minutes of arc to add to those of the whole hour of GHA of the sun.

3. The addition of (1) and (2) gives the GHA of the sun in degrees, minutes, and tenths of minutes of arc for the exact hour, minute, and second of your sextant shot. In west longitude this corresponds directly to your vessel's longitude. In east longitude, GHA is subtracted from 360° to obtain your vessel's longitude.

F. The sextant angle at apparent local noon, selected in step C above, serves as the basis for determining latitude as follows:

1. Take raw sextant reading or sextant altitude (Hs).   Hs

2. Adjust for sextant index error by making the appropriate index correction (IC). $\underline{+/-IC}$

3. The result is corrected Hs.   Corr Hs

4. Subtract dip for height of eye from inside front cover of the N.A. $\underline{-dip}$

5. The result is apparent altitude (Ha).   Ha

6. From the inside cover of the N.A. under "Sun," select month period and upper or lower limb correction at, or immediately above, the apparent altitude listed. This is the altitude correction (Alt. Corrn) factor for several items and is added or subtracted, as indicated.   +/−Alt. Corrn

7. The result is observed altitude (Ho) of the sun.   Ho

8. Write down zenith distance (zd), which is 90°, as: 89°60.0′

9. Subtract observed altitude.   −Ho

10. The result is co-alti-
    tude.

    Co-Alt.

11. Return to the white
    pages of the N.A. for
    the same date and
    hour as in E.1 above.
    Under the "Sun" col-
    umn is Dec (declina-
    tion). From approx-
    imately March 21 to
    September 21, Dec is
    north (N), and be-
    tween September 21
    and March 21, Dec is
    south (S). Whole de-
    grees are stated at
    each six hours with
    minutes and tenths of
    minutes shown for
    each hour. If Dec is
    whichever you are
    (north or south of the
    equator), then add; if
    Dec is of a different
    hemisphere from you
    (i.e., you are north of
    equator, the sun is
    south), then you will

subtract from co-al-
titude. As Dec in-
creases or decreases
so slowly, visual in-
spection and inter-
polation   between
hours of GMT will in-
dicate how much, if
any, Dec should be
adjusted for the time
after the whole hour. $+/-$Dec

12. The result is your lati-
     tude.   Latitude

G. Compare calculated longitude and latitude
to your original DR position estimates to
check for reasonableness. An examination of
the steps followed should resolve significant
differences.

H. The "noon" reduction described above
was for a sighting of the sun, since that is the
most commonly used celestial body. However, the concept applies to all bodies crossing the meridian of longitude of the navigator.
The moment of greatest observed altitude is
the time of "noon." In practice, the moon is
the second most commonly used body for

"noon" sightings. Planets may also be used, though stars are seldom used due to the slowness of their apparent progression across the skies and difficulty in confirming the horizon.

4.

SUN LINE
OF POSITION

Sun Line of Position

Items needed: Sextant, *The Nautical Almanac* (N.A.), *Sight Reduction Tables* (H.O. 249, Vol. II or III), accurate time, plotting paper, graph paper, straightedge, scratch paper, pencil, dividers, and plastic plotter.

A. Note body (sun), date, and deduced latitude and longitude (DR position). Select the upper or lower portion (limb) of the sun to bring down to the horizon. Determine height of eye above water.

B. Depending on skill, experience, and conditions, take a number of sextant shots, recording exact watch time of each.

C. To select the desired sextant reading and time, proceed as follows: On graph paper, place on left side a vertical scale in degrees and minutes of arc for sextant readings. Across the bottom, place a horizontal scale for time. Arrange, or plot, sextant readings as recorded for degrees and time. With a straightedge, place a line through the apparent *flow* of the plotted points. Select the raw sextant reading and watch time nearest the drawn line. These are the shot and time to be used.

D. It is necessary to make certain corrections and adjustments to the raw sextant reading selected in C above, as follows:

1. Take raw sextant reading (Hs).   Hs

2. Adjust for sextant index error.   +/−IC

3. The result is corrected Hs.   Corr Hs

4. Subtract dip for height of eye from inside front cover of the N.A.   −dip

5. The result is apparent
   altitude.   Ha

6. From the inside front
   cover of the N.A. un-
   der "Sun," select
   month period and up-
   per or lower limb
   correction at, or im-
   mediately above, the
   apparent altitude
   listed. This is the altitude correction factor
   for several items and
   is added or sub-
   tracted, as indicated.   +/−Alt. Corrn

7. The result is the ob-
   served altitude of the
   sun.   Ho

E. Calculate Greenwich mean time (GMT)
   from the time of the selected sextant shot as
   follows:

1. Take raw watch time (WT) in hours,
   minutes, and seconds.

2. Correct for watch error (WE): subtract for fast, add for slow.

3. Confirm A.M. versus P.M. If later than 1:00 P.M., add 12 hours.

4. The sum of (1), (2), and (3) = ZT (zone time).

5. Add zone description (ZD) to ZT to get GMT. For example, Miami, Florida, is in Zone 5, as zones are 15° wide from Greenwich (80° divided by 15° = 5+ hours); however, during summer, if daylight saving time (DST) is in effect, add one less hour (i.e., 5 − 1 = 4 hours) to ZT to get GMT.

6. If GMT as calculated is greater than 24 hours, subtract 24 hours from time and add one day to get new Greenwich date; the remainder of the GMT becomes the new time.

F. Calculate the Greenwich hour angle (GHA) of the sun as follows:

1. Turn to the proper Greenwich date in the white pages of the N.A. Under the column for the sun locate the GHA for the whole hour of GMT. Write down the degrees, minutes, and tenths of minutes shown.

2. To interpolate the GHA for the remaining minutes and seconds of GMT, turn to the yellow pages entitled "Increments and Corrections" in the back of the N.A. These pages are marked in minutes of time with seconds of time running down the left-hand columns. Two minutes, with 60 seconds each, are displayed on a page. Having selected the correct minute and second of time, take from the "Sun/Planets" column the degrees, minutes, and tenths of minutes of arc to add to those of the whole hour of GHA of the sun.

3. The addition of (1) and (2) gives the GHA of the sun in degrees, minutes, and tenths of minutes of arc for the exact hour, minute, and second of the sextant shot. In west longitude this corresponds to the geographical

longitude of the sun. In east longitude, GHA is subtracted from 360° to obtain the sun's longitude.

G. Calculate latitude of the sun as follows:

1. Return to the white pages of the N.A. for the same date and hour as in F.1 above. Under the "Sun" column is Dec (declination). From approximately March 21 to September 21, Dec is north (N), and between September 21 and March 21, Dec is south (S). Whole degrees are indicated at each six hours with minutes and tenths shown for each hour. At the bottom of that column is a little *d* that indicates the hourly change in the Dec. A visual interpolation for the actual GMT will produce a factor to be added to, or subtracted from, Dec (depending on the apparent increase or decrease in Dec as seen in the table) to complete the Dec calculation. This corresponds to the geographic latitude of the sun.

2. The determination of the GHA and the Dec of the sun provides the geo-

graphic position (GP) of the sun. It is from this location that your line of position (LOP) is calculated.

H. For purposes of computation and to choose the appropriate tables, it is necessary to assume a position. This is not your DR position; it is merely one nearby that facilitates the use of tables and avoids complex calculations. The initial steps are as follows:

1. Select the latitude in whole degrees that is nearest your DR latitude. This is the assumed latitude.

2. In west longitude select the nearest longitude that would end with exactly as many minutes and tenths of minutes as the GHA of the sun when you took the sight (F.2 above). When in east longitude, change the minutes and tenths of minutes of your DR longitude to equal the difference between the GHA and a whole degree so that, when adding, the assumed longitude comes out to a whole degree.

3. Plotting (1) and (2) on plotting paper gives your assumed position (AP).

I. From the GHA calculated in F, and with the assumed longitude determined in H.2 above, it is possible to construct your local hour angle (LHA), which is the angle measured in a westward direction only from your AP to the meridian of the sun. It is essential to accept that when your vessel is in west longitude, LHA is less than GHA because the sun must pass Greenwich before it passes you. Conversely, when your vessel is in east longitude, LHA is greater than GHA since the sun passes you before it reaches Greenwich. The LHA is calculated as follows:

1. When in west longitude, if your assumed longitude is less than the GHA, the LHA is obtained by subtracting the assumed longitude from the GHA of the sun. However, if your assumed longitude is greater than the GHA, the LHA is obtained by *first* adding 360° to the GHA and *then* subtracting your assumed longitude.

2. When in east longitude, the LHA

is obtained by simply adding the assumed longitude to the GHA of the sun. If the total of your assumed longitude and the GHA exceeds 360°, the LHA is then obtained by subtracting 360°.

3. It should be noted that the assumed longitude was determined to be a meridian of longitude that caused the resulting LHA to be in whole degrees.

J. With the Dec of the sun from G, the assumed latitude from H, and the LHA from I, it is now possible to enter the tables of H.O. 249, Vol. II, and pursue the material for the line of position (LOP). Note: For latitudes of 40° or more, H.O. 249, Vol. III, is comparably arranged and similarly used.

1. In the tables for latitudes from 0° to 39°, for each whole degree of latitude there are four headings: "Declination (0°-14°) *Same* Name As Latitude," "Declination (0°-14°) *Contrary* Name to Latitude," and two similar headings for declinations

from 15° to 29°. *"Same* Name" or *"Contrary* Name" is determined by whether the north or south declination of the sun puts it in the same hemisphere as your vessel. Therefore, when sailing in the Northern Hemisphere, you use the *"Same* Name" from late March to September and *"Contrary* Name" from September to March.

2. Turn to the table section that gives information for your assumed latitude.

3. Within the section of the assumed latitude, turn to the page that gives figures that relate to the sun's declination in whole degrees.

4. Within the assumed latitude section for the sun's declination in degrees, select the subsection that corresponds to the *same* or *contrary* name of the declination, depending on the hemispheres of you and the sun.

K. Having located the proper page in H.O.

249, Vol. II, which contains the assumed latitude, the correct declination in whole degrees, and the appropriate name of declination, you can obtain the final pieces of information.

1. Down the extreme left and right sides of the page are columns entitled "LHA." Proceed down these columns until you locate the one for your LHA (which came from I above). At the intersection of the appropriate LHA and the column of the declination in whole degrees is a series of numbers. The first set has four places and at the top of the column is labeled "Hc" (computed altitude) in degrees and minutes. The second set has one or two places and is labeled "d" and is in minutes. The third set has two or three places and is labeled "Z" and is in whole degrees.

2. Note carefully the degrees and minutes under "Hc." This is your computed altitude, assuming that the sun's declination is in whole degrees, which it almost never is. Therefore, you must interpolate for that portion

of declination which exceeded the whole degrees of declination. The number under "d" is the key to the interpolation to get a completed Hc.

3. Note the plus (+) or minus (−) sign at, or just above, the numbers under "d" as well as the numbers themselves. Also note the remaining minutes of declination not accounted for. Staying within Vol. II, turn to page 242, Table 5, "Correction to Tabulated Altitude for Minutes of Declination." The row across the top ranges from 1 to 60 and is used for "d." Locate the line of numbers for "d." At the edges of the table are two identical columns running from 0 to 59 and headed with the sign for minutes ('). Proceed down either column to the number of minutes of declination that need to be interpolated. At the intersection of the two ("d" and minutes) is the value in minutes for the correction to Hc. (In truth, accidental reversal of "d" and minutes will make no difference, as the answer is the same.) This number of minutes should be added to, or sub-

tracted from, Hc depending on the plus or minus sign previously noted for "d." The sum of Hc in whole degrees and minutes for "d" from Table 5 is the complete Hc for your sight.

4. At this point almost the only remaining step prior to plotting the LOP is to determine the direction from the AP. This is accomplished with the value for Z in degrees noted above. For all northern latitudes in the top left-hand corner of the pages of Vol. II is the legend, "LHA greater than 180°, Zn = Z; LHA less than 180°, Zn = 360 − Z." Therefore, if your LHA is greater than 180°, do nothing. If your LHA is less than 180°, subtract the value of Z from 360°. In both cases the result is Zn, or azimuth. Instructions for sailing in southern latitudes are in the bottom left-hand corner.

5. Three critical items of information have been developed in this section. The first two result from a comparison of Hc and Ho. One comparison provides the direction toward or

away from the sun you will place the
LOP in relation to the AP. The other
comparison provides the distance
the LOP will be from the AP. The
third piece of information is the Zn,
which is the exact direction in whole
degrees from the AP toward which
the sun is located.

L. The final phase of this process is determining the direction, distance, and placement of
the LOP in relation to the AP. This is accomplished as follows:

1. Compare the Ho and the Hc. If the
Ho is larger than the Hc, then the
LOP will be placed toward the sun. If
the Ho is less than the Hc, then the
LOP will be placed away from the
sun.

2. The LOP is a line drawn perpendicular to the azimuth line. The distance
along the azimuth is the intercept
(Int). The Int is the difference between the Ho and the Hc. This difference in minutes of arc corresponds
directly to the distance in nautical

miles that the LOP will be drawn across the azimuth from the AP.

3. As indicated, the azimuth (Zn) is the direction in degrees toward which the sun lies, and along which you draw a line through your AP.

4. With pencil and plotter draw a line through the AP along the Zn, extending away from the sun if the Hc is larger than the Ho, and toward the sun if the Ho is greater than the Hc. Then take the dividers and measure the distance along the longitude scale of the chart or plotting paper of the Int, which was the difference in minutes between the Hc and Ho. Marking from the AP, draw a line perpendicular to the azimuth at the point of intercept. This perpendicular line is the LOP. Your vessel should be located approximately along this LOP. The intersection of multiple LOPs from other bodies constitutes a fix.

M. Compare your LOP or fix with your DR position. Unless you have made a mistake in the calculation of either the LOP or the DR position, the two should be reasonably close, considering the circumstances. An examination of the steps should resolve significant differences.

5.

MOON LINE
OF POSITION

Moon Line of Position

Items needed: Sextant, *The Nautical Almanac* (N.A.), *Sight Reduction Tables* (H.O. 249, Vol. II or III), accurate time, plotting paper, graph paper, straightedge, scratch paper, pencil, dividers, and plastic plotter.

A. Note body (moon), date, deduced latitude and longitude (DR position). Select the upper or lower portion (limb) of the moon to bring down to the horizon. Determine height of eye above water.

B. Depending on skill, experience, and conditions, take a number of sextant shots, recording exact watch time of each.

C. To select the desired sextant reading and time, proceed as follows: On graph paper, place on the left side a vertical scale in degrees and minutes of arc for sextant readings. Across the bottom, place a horizontal scale for time. Arrange, or plot, sextant readings as recorded for degrees and time. With a straightedge, place a line through the apparent *flow* of the plotted points. Select the raw sextant reading and watch time nearest the drawn line. These are the shot and time to be used.

D. Calculate Greenwich mean time (GMT) from the time of the selected sextant shot as follows:

1. Take raw watch time (WT) in hours, minutes, and seconds.

2. Correct for watch error (WE): subtract for fast, add for slow.

3. Confirm A.M. versus P.M. If later than 1:00 P.M., add 12 hours.

4. The sum of (1), (2), and (3) = ZT (zone time).

5. Add zone description (ZD) to ZT to get GMT. For example, Miami, Florida, is in Zone 5, as zones are 15° wide from Greenwich (80° divided by 15° = 5+ hours); however, during summer, if daylight saving time (DST) is in effect, add one less hour (i.e., 5 − 1 = 4 hours) to ZT to get GMT.

6. If GMT as calculated is greater than 24 hours, subtract 24 hours from time and add one day to get new Greenwich date; the remainder of the GMT becomes the new time.

E. Calculate the Greenwich hour angle (GHA) of the moon as follows:

1. Turn to the proper Greenwich date in the white pages of the N.A. Under the column for the moon, locate the GHA for the whole hour of GMT. Write down the degrees, minutes, and tenths of minutes shown.

2. Next to the GHA in whole hours is a column under the letter "v" in minutes and tenths of minutes. This is a

secondary correction of the GHA adjustment for GMT time past the whole hour. Note the "v" factor; it will be used shortly.

3. To interpolate the GHA for the remaining minutes and seconds of GMT, turn to the yellow pages entitled "Increments and Corrections" in the back of the N.A. These pages are marked in minutes of time with seconds of time running down the left-hand columns. Two minutes, with 60 seconds each, are displayed on a page. Having selected the correct minute and second of time, take from the "Moon" column the degrees, minutes, and tenths of minutes of arc to add to those of the whole hour of GHA of the moon.

4. Remaining in the same minute on the page of "Increments and Corrections," note three adjoining columns labeled "v or d Corrn." Under the portion headed by the "v," find the same number noted in (2) above; to its right is a figure under the portion headed by "Corrn." This is the cor-

rection factor for "v." It is to be added to the GHA to complete the calculation.

5. The addition of (1), (3), and (4) gives the GHA of the moon in degrees, minutes, and tenths of minutes of arc for the exact hour, minute, and second of time of the sextant shot. In west longitude this corresponds to the geographical longitude of the moon. In east longitude, GHA is subtracted from 360° to obtain the moon's longitude.

F. Calculate the latitude of the moon as follows:

1. Return to the white pages of the N.A. for the same date and hour as in E.1 above. Under the "Moon" section is a column for Dec (declination). Sometimes the moon is north (N) of the equator, at other times south (S). Thus, Dec may be N or S and is so indicated at each six hours of GMT. For the GMT in whole hours note whether Dec is N or S and the degrees, minutes, and tenths of minutes

of its declination. Visually observe whether Dec is increasing or decreasing at the following whole hour of GMT. This determines whether the correction factors for Dec are to be added or subtracted.

2. To the right of Dec is a column headed with a "d" in minutes and tenths of minutes. This is the only additional correction for Dec past the whole hour of GMT. Note the "d" factor; it will be used in the next step.

3. Return to the yellow pages of the N.A. entitled "Increments and Corrections" for the minutes of time remaining beyond the full hour of GMT (this is the same page as for E.3 and E.4 above). Under the columns headed "v or d Corrn" for the appropriate minute of time, locate under the portion headed by "d" the same number noted in (2) above; to its right is a figure under the portion headed by "Corrn." This is the correction factor for "d." It is to be added to, or subtracted from, Dec (depending on whether declination

is increasing or decreasing) to complete the Dec calculation.

4.   The determination of Dec in degrees, minutes, and tenths of minutes of arc, and the notation as to whether it is N or S, for the exact hour and minute of time of the sextant shot, produce the geographical latitude of the moon.

5.   The determination of the GHA and Dec of the moon provides the geographic position (GP) of the moon. It is from this location that your line of position (LOP) is calculated.

G.   At this point in the process to determine your line of position, it is necessary to return to the white pages of the N.A. Returning to the correct date and whole hour of GMT under the section entitled "Moon" is in E above, locate to the right of "d" a column headed "H.P." (horizontal parallax). This is a large number in the neighborhood of 60 minutes of arc and is expressed in minutes and tenths of minutes. Note this "H.P." number, for you will need it momentarily.

H. It is necessary to make certain corrections and adjustments to the raw sextant reading selected in C above as follows:

1. Take raw sextant reading (Hs).   Hs

2. Adjust for sextant index error.   + / –IC

3. The result is corrected Hs.   Corr Hs

4. Subtract dip for height of eye from inside front cover of the N.A.   –dip

5. The result is apparent altitude.   Ha

6. From the last two pages of the N.A., "Altitude Correction Tables for Moon," select from the upper part of the table the correction closest to your Ha. The Main Correction is for re-

fraction and is taken by first entering the table headed with whole degrees of "App. Alt." in 5-degree increments. Proceed down this column until you locate the whole degree of your Ha. Along the sides of the tables on both pages are minutes of "App. Alt." in 10-minute increments. Now enter the table from the minutes column on the side corresponding to the number of minutes necessary. No calculations are required, as the correction factor is readily apparent. Any interpolation is minimal and may be done by eye or ignored. The total correction is in minutes and tenths of

minutes of arc and is
added to the Ha.   <u>+ Main Corrn</u>

7. The second correc-
tion is for horizontal
parallax (H.P.) and is
taken from the lower
part of the same table,
and from the same
column as the Main
Correction. Select the
lower (L) or upper (U)
limb correction (as
shot by sextant), cor-
responding to the
horizontal parallax
(H.P.) factor note in G
above, and add.   <u>+ Limb Corrn</u>

8. If upper limb sextant
shot is used, subtract
30′.   <u>−If upper</u>

9. The result is the ob-
served altitude of the
moon.   Ho

I. For purposes of computation and to choose
the appropriate tables, it is necessary to as-

sume a position. This is not your DR position; it is merely one nearby that facilitates the use of tables and avoids complex calculations. The initial steps are as follows:

1. Select the latitude in whole degrees that is nearest your DR latitude. This is the assumed latitude.

2. In west longitude select the nearest longitude that would end with exactly as many minutes and tenths of minutes as the GHA of the moon when you took the sight (E.5 above). When in east longitude, change the minutes and tenths of minutes of your DR longitude to equal the difference between the GHA and a whole degree so that, when adding, the assumed longitude comes out to a whole degree.

3. Plotting (1) and (2) on plotting paper gives your assumed position (AP).

J. From the GHA calculated in E and with the assumed longitude determined in I.2 above, it is possible to construct your local hour angle (LHA), which is the angle measured in a westward direction only from your AP to the meridian of the moon. It is essential to accept that when your vessel is in west longitude, LHA is less than GHA because the moon must pass Greenwich before it passes you. Conversely, when your vessel is in east longitude, LHA is greater than GHA since the moon passes you before it reaches Greenwich. The LHA is calculated as follows:

1. When in west longitude, if your assumed longitude is less than the GHA, the LHA is obtained by subtracting the assumed longitude from the GHA of the moon. However, if your assumed longitude is greater than the GHA, the LHA is obtained by *first* adding 360° to the GHA and *then* subtracting your assumed longitude.

2. When in east longitude, the LHA is obtained by simply adding the assumed longitude to the GHA of the moon. If the total of your assumed longitude and the GHA exceeds 360°, the LHA is then obtained by subtracting 360°.

3. It should be noted that the assumed longitude was determined to be a meridian of longitude that caused the resulting LHA to be in whole degrees.

K. With the Dec of the moon from F, the assumed latitude from I, and the LHA from J, it is now possible to enter the tables of H.O. 249, Vol. II, and pursue the material for the line of position (LOP). Note: For latitudes of 40° or more, H.O. 249, Vol. III, is comparably arranged and similarly used.

1. In the tables for latitudes from 0° to 39°, for each whole degree of latitude there are four headings: "Declination (0°-14°) *Same* Name As Latitude," "Declination (0°-14°) *Contrary* Name to Latitude," and two similar headings for declinations from 15° to 29°. *"Same* Name" or *"Contrary* Name" is determined by whether the north or south declination of the moon puts it in the same hemisphere as your vessel.

2. Turn to the table section that gives

information for your assumed latitude.

3. Within the section of the assumed
    latitude, turn to the page that gives
    figures relating to the moon's decli-
    nation in whole degrees.

4. Within the assumed latitude section
    for the moon's declination in de-
    grees, select the subsection that cor-
    responds to the *same* or *contrary*
    name of the declination, depending
    on the hemispheres of you and the
    moon.

L. Having located the proper page in H.O. 249,
    Vol. II, which contains the assumed latitude,
    the correct declination in whole degrees, and
    the appropriate name of declination, you can
    obtain the final pieces of information.

1. Down the extreme left and right
    sides of the page are columns en-
    titled "LHA." Proceed down these
    columns until you locate the one for
    your LHA (which came from J
    above). At the intersection of the ap-
    propriate LHA and the column of the

declination in whole degrees is a series of numbers. The first set has four
places and at the top of the column is
labeled "Hc" (computed altitude) in
degrees and minutes. The second set
has one or two places and is labeled
"d" and is in minutes. The third set
has two or three places and is labeled
"Z" and is in whole degrees.

2.   Note carefully the degrees and minutes under "Hc." This is your computed altitude, assuming that the
moon's declination is in whole degrees, which it almost never is.
Therefore, you must interpolate for
that portion of declination which exceeded the whole degrees of declination. The number under "d" is the
key to the interpolation to get a completed Hc.

3.   Note the plus (+) or minus (−) sign
at, or just above, the numbers under
"d" as well as the numbers themselves. Also note the remaining minutes of declination not accounted
for. Staying within Vol. II, turn to
page 242, Table 5, "Correction to

Tabulated Altitude for Minutes of Declination." The row across the top ranges from 1 to 60 and is used for "d." Locate the line of numbers for "d." At the edges of the table are two identical columns running from 0 to 59 and headed with the sign for minutes ('). Proceed down either column to the number of minutes of declination that need to be interpolated. At the intersection of the two ("d" and minutes) is the value in minutes for the correction to Hc. (In truth, accidental reversal of "d" and minutes will make no difference, as the answer is the same.) This number of minutes should be added to, or subtracted from, Hc depending on the plus or minus sign previously noted for "d." The sum of Hc in whole degrees and minutes for "d" from Table 5 is the complete Hc for your sight.

4. At this point almost the only remaining step prior to plotting the LOP is to determine the direction from the AP. This is accomplished with the value for Z in degrees noted above. For all northern latitudes in the top

left-hand corner of the pages of Vol. II is the legend, "LHA greater than 180°, Zn = Z; LHA less than 180°, Zn = 360 − Z." Therefore, if your LHA is greater than 180°, do nothing. If your LHA is less than 180°, subtract the value of Z from 360°. In both cases the result is Zn, or azimuth. Instructions for sailing in southern latitudes are in the bottom left-hand corner.

5. Three critical items of information have been developed in this section. The first two result from a comparison of Hc and Ho. One comparison provides the direction toward or away from the moon you will place the LOP in relation to the AP. The other comparison provides the distance the LOP will be from the AP. The third item of information is the Zn, which is the exact direction in whole degrees from the AP toward which the moon is located.

M. The final phase of this process is determining the direction, distance, and placement of the LOP in relation to the AP. This is accomplished as follows:

Moon Line of Position

1. Compare the Ho and the Hc. If the Ho is larger than the Hc, then the LOP will be placed toward the moon. If the Ho is less than the Hc, then the LOP will be placed away from the moon.

2. The LOP is a line drawn perpendicular to the azimuth line. The distance along the azimuth is the intercept (Int). The Int is the difference between the Ho and the Hc. This difference in minutes of arc corresponds directly to the distance in nautical miles that the LOP will be drawn across the azimuth from the AP.

3. As indicated, the azimuth (Zn) is the direction in degrees toward which the moon lies, and along which you draw a line through your AP.

4. With pencil and plotter, draw a line through the AP along the Zn, extending away from the moon if the Hc is larger than the Ho, and toward the moon if the Ho is greater than the Hc. Then take the dividers and measure the distance along the longitude

scale of the chart or plotting paper of the Int, which was the difference in minutes between the Hc and Ho. Marking from the AP, draw a line perpendicular to the azimuth at the point of intercept. This perpendicular line is the LOP. Your vessel should be located approximately along this LOP. The intersection of multiple LOPs from other bodies constitutes a fix.

N. Compare your LOP or fix with your DR position. Unless you have made a mistake in the calculation of either the LOP or the DR position, the two should be reasonably close, considering the circumstances. An examination of the steps should resolve significant differences.

6.

POLARIS LATITUDE
LINE OF POSITION

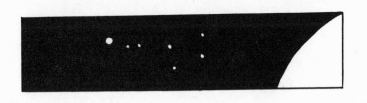

Polaris Latitude Line of Position

Items needed: Sextant, *The Nautical Almanac* (N.A.), accurate time, graph paper, straightedge, scratch paper, and pencil.

A. Note body (Polaris), date, deduced latitude and longitude (DR position). Determine height of eye above water.

B. Depending on skill, experience, and conditions, take a number of sextant shots, recording exact watch time of each.

C. To select the desired sextant reading and time, proceed as follows: On graph paper, place on the left side a vertical scale in degrees and minutes of arc for sextant readings. Across the bottom, place a horizontal scale for time. Arrange, or plot, sextant readings as recorded for degrees and time. With a straightedge, place a line through the apparent *flow* of the plotted points. Select the raw sextant reading and watch time nearest the drawn line. These are the shot and time to be used.

D. Calculate Greenwich mean time (GMT) from the time of the selected sextant shot as follows:

1. Take raw watch time (WT) in hours, minutes, seconds.

2. Correct for watch error (WE): subtract for fast, add for slow.

3. Confirm A.M. versus P.M. If later than 1:00 P.M., add 12 hours.

4. The sum of (1), (2), and (3) = ZT (zone time).

5. Add zone description (ZD) to ZT to get GMT. For example, Miami, Florida, is in Zone 5, as zones are 15°

wide from Greenwich (80° divided by 15° = 5+ hours); however, during summer, if daylight saving time (DST) is in effect, add one less hour (i.e., 5 − 1 = 4 hours) to ZT to get GMT.

6. If GMT as calculated is greater than 24 hours, subtract 24 hours from time and add one day to get new Greenwich date; the remainder of the GMT becomes the new time.

E. Calculate local hour angle (LHA) of Aries as follows:

1. Turn to the proper Greenwich date in the white pages of the N.A. Under the left-hand GMT column, entitled "Aries," locate the GHA of Aries for the whole hour of GMT. Write down the degrees, minutes, and tenths of minutes shown.

2. To interpolate the GHA of Aries for the remaining minutes and seconds of GMT, turn to the yellow pages entitled "Increments and Corrections" in the back of the N.A. These pages

are marked in minutes of time with seconds of time running down the left-hand columns. Two minutes, with 60 seconds each, are displayed on a page. Having taken the correct minute and second of time, take from the "Aries" column the degrees, minutes, and tenths of minutes of arc to add to those of the whole hour of GHA of Aries.

3. The addition of (1) and (2) gives the GHA of Aries in degrees, minutes, and tenths of minutes of arc for the exact hour, minute, and second of time of the sextant shot. When in west longitude, if your estimated longitude is less than the GHA of Aries, the LHA is obtained by subtracting your longitude from the GHA of Aries. However, if your assumed longitude is greater than the GHA of Aries, the LHA is obtained by *first* adding 360° to the GHA of Aries and *then* subtracting your assumed longitude.

4. Conversely, when in east longitude, the LHA is obtained by simply add-

ing your estimated longitude to the GHA of Aries. If the total of your assumed longitude and the GHA of Aries exceeds 360°, the LHA is then obtained by subtracting 360°.

F. Take the LHA of Aries and turn to the last three pages in the white pages of the N.A. entitled "Polaris (Pole Star) Tables." Several factors are obtained under the column of the LHA of Aries, which is grouped in increments of 10°. Select the column of the calculated LHA of Aries (E.4 above). The factors given are labeled a_0, a_1, and a_2, for reference only. The azimuth of Polaris is also listed. The addition of a_0, a_1, and a_2 gives a total correction and will be part of a final adjustment of the observed sextant angle (Ho) below.

G. To calculate latitude it is necessary to make certain corrections to the raw sextant reading selected in C above, as follows:

1. Take raw sextant reading (Hs).   Hs

2. Adjust for sextant index error. $+/-IC$

3. The result is corrected
   Hs.   Corr Hs

4. Subtract dip for height
   of eye from inside
   front cover of the N.A. $-$dip

5. The result is apparent
   altitude.   Ha

6. From the inside cover
   of the N.A. under
   "Stars and Planets,"
   select correction for
   refraction at, or im-
   mediately above, the
   apparent altitude
   listed. This is the altitude correction factor
   and is always sub-
   tracted. $-$Alt. Corrn

7. The result is the ob-
   served altitude of Po-
   laris.   Ho

8. Subtract one whole
   degree to provide for
   base from which a_0,

a$_1$, and a$_2$ were
calculated. $-1°00.0'$

9. Add the total from F
   above.   + F above

10. The result is your lati-
    tude.   Latitude

H. Compare calculated latitude from original
estimate of latitude to check for reasonableness. An examination of the steps should resolve significant differences.

7.

PLANET LINE
OF POSITION

Planet Line of Position

Items needed: Sextant, *The Nautical Almanac* (N.A.), *Sight Reduction Tables* (H.O. 249, Vol. II and III), accurate time, plotting paper, graph paper, straightedge, scratch paper, pencil, dividers, and plastic plotter.

A. Note body (planet), date, deduced latitude and longitude (DR position). Determine height of eye above water.

B. Depending on skill, experience, and conditions, take a number of sextant shots, recording exact watch time of each.

C. To select the desired sextant reading and

time, proceed as follows: On graph paper, place on the left side a vertical scale in degrees and minutes of arc for sextant readings. Across the bottom, place a horizontal scale for time. Arrange, or plot, sextant readings as recorded for degrees and time. With a straightedge, place a line through the apparent *flow* of the plotted points. Select the raw sextant reading and watch time nearest the drawn line. These are the shot and time to be used.

D. It is necessary to make certain corrections and adjustments to the raw sextant reading selected in C above, as follows:

1. Take raw sextant reading (Hs).   Hs

2. Adjust for sextant index error.   + / −IC

3. The result is corrected Hs.   Corr Hs

4. Subtract dip for height of eye from inside front cover of the N.A.   −dip

5. The result is apparent altitude.   Ha

6. From the inside front cover of the N.A. under "Stars and Planets," select the correction for refraction at, or immediately above, the apparent altitude listed. This is the main altitude correction for all four planets commonly used and is always subtracted.

    —Alt. Corrn

7. Remaining at the same place in the N.A., note another column for additional corrections for two planets, Venus and Mars. If shooting either of these, select the correction factor that applies for your seasonal or monthly period as it relates to

the Ha. Observe that this factor is small, measuring in only tenths of minutes of arc. The additional correction is always added (except when Venus has a higher altitude than the sun, as described in the N.A.).

+ Add'l Corrn

8. The result is the observed altitude of the planet.

Ho

E. Calculate Greenwich mean time (GMT) from the time of the selected sextant shot as follows:

1. Take raw watch time (WT) in hours, minutes, and seconds.

2. Correct for watch error (WE): subtract for fast, add for slow.

3. Confirm A.M. versus P.M. If later than 1:00 P.M., add 12 hours.

4. The sum of (1), (2), and (3) = ZT (zone time).

5. Add zone description (ZD) to ZT to get GMT. For example, Miami, Florida, is in Zone 5, as zones are 15° wide from Greenwich (80° divided by 15° = 5+ hours); however, during summer, if daylight saving time (DST) is in effect, add one less hour (i.e., 5 − 1 = 4 hours) to ZT to get GMT.

6. If GMT as calculated is greater than 24 hours, subtract 24 hours from time and add one day to get new Greenwich date; the remainder of the GMT becomes the new time.

F. Calculate the Greenwich hour angle (GHA) of the planet as follows:

1. Turn to the proper Greenwich date in the white pages of the N.A. On the left side page are four columns. The planets listed are Venus, Mars, Jupiter, and Saturn. They are the four commonly used in celestial navigation. To the right of the name of each

planet is a plus (+) or minus (−) figure; this relates to the magnitude or
brightness of the planet. Venus is the
brightest planet, Jupiter the second
brightest. Mars and Saturn may vary
at times. In any event, the largest
negative number is the brightest and
the largest positive number is the
least bright. Under the column for
the planet, locate the GHA for the
whole hour of GMT. Write down the
degrees, minutes, and tenths of minutes shown.

2. At the bottom of the planet column
    is a "v" in minutes and tenths of min-
    utes. This is a secondary correction
    of the GHA adjustment for GMT
    time past the whole hour. Note the
    "v" factor; it will be used shortly.
    Also observe that Venus is often
    negative while the other three are
    positive.

3. To interpolate the GHA for the re-
    maining minutes and seconds of
    GMT, turn to the yellow pages en-
    titled "Increments and Corrections"
    in the back of the N.A. These pages

are marked in minutes of time with seconds of time running down the left-hand columns. Two minutes, with 60 seconds each, are displayed on a page. Having selected the correct minute and second of time, take from the "Sun and Planets" column the degrees, minutes, and tenths of minutes of arc to add to those of the whole hour of GHA of the planet.

4. Remaining in the same minute on the page of "Increments and Corrections," note three adjoining columns labeled "v or d Corrn." Under the portion headed by the "v," find the same number noted in (2) above; to its right is a figure under the portion headed by "Corrn." This is the correction factor for "v." It is to be added to the GHA in the cases of Mars, Jupiter, and Saturn, and subtracted from the GHA in the case of Venus when the sign is negative, to complete the calculation.

5. The addition of (1), (3), and (4) gives the GHA of the planet in degrees, minutes, and tenths of minutes of arc

for the exact hour, minute, and second of time of the sextant shot. In west longitude this corresponds to the geographic longitude of the planet. In east longitude, GHA is subtracted from 360° to obtain the planet's longitude.

G. Calculate the latitude of the planet as follows:

1. Return to the white pages of the N.A. for the same date and hour as in F.1 above. Under the appropriate "Planet" column is Dec (Declination). At times the planet is north (N) of the equator, at other times south (S). Thus, Dec may be N or S. Whole degrees are indicated at each six hours with minutes and tenths of minutes shown each hour. At the bottom of the planet column is a "d," which indicates the hourly change in Dec. A visual interpolation for the actual GMT will produce a factor to be added to, or subtracted from, Dec (depending on the apparent increase or decrease in Dec as seen in the table), which will complete the Dec

calculation. This corresponds to the geographic latitude of the planet.

2. The determination of the GHA and the Dec of the planet provides the geographic position (GP) of the planet. It is from this location that your line of position (LOP) is calculated.

H. For purposes of computation and to select the appropriate tables, it is necessary to assume a position. This is not your DR position; it is merely one nearby that facilitates the use of tables and avoids complex calculations. The initial steps are as follows:

1. Select the latitude in whole degrees that is nearest your DR latitude. This is the assumed latitude.

2. In west longitude select the nearest longitude that would end with exactly as many minutes and tenths of minutes as the GHA of the planet when you took the sight (F.5 above). When in east longitude, change the minutes and tenths of minutes of your DR longitude to equal the dif-

ference between the GHA and a whole degree so that, when adding, the assumed position comes out to a whole degree.

3. Plotting (1) and (2) on plotting paper gives your assumed position (AP).

I. From the GHA calculated in F, and with the assumed longitude determined in H.2 above, it is possible to construct your local hour angle (LHA), which is the angle measured in a westward direction only from your AP to the meridian of the planet. It is essential to accept that when your vessel is in west longitude, LHA is less than GHA because the planet must pass Greenwich before it passes you. Conversely, when your vessel is in east longitude, LHA is greater than GHA since the planet passes you before it reaches Greenwich. The LHA is calculated as follows:

1. When in west longitude, if your assumed longitude is less than the GHA, the LHA is obtained by subtracting the assumed longitude from the GHA of the planet. However, if your assumed longitude is greater than the GHA, the LHA is obtained

by *first* adding 360° to the GHA and *then* subtracting your assumed longitude.

2. When in east longitude, the LHA is obtained by simply adding the assumed longitude to the GHA of the planet. If the total of your assumed longitude and the GHA exceeds 360°, the LHA is then obtained by subtracting 360°.

3. It should be noted that the assumed longitude was determined to be a meridian of longitude that caused the resulting LHA to be in whole degrees.

J. With the Dec of the planet from G, the assumed latitude from H, and the LHA from I, it is now possible to enter the tables of H.O. 249, Vol. II, and pursue the material for the line of position (LOP). Note: For latitudes of 40° or more, H.O. 249, Vol. III, is comparably arranged and similarly used.

1. In the tables for latitudes from 0° to 39°, for each whole degree of lati-

tude there are four headings: "Declination (0°-14°) *Same* Name As Latitude," "Declination (0°-14°) *Contrary* Name to Latitude," and two
similar headings for declinations
from 15° to 29°. *"Same* Name" or
"Contrary Name" is determined by
whether the north or south declination of the planet puts it in the same
hemisphere as your vessel.

2. Turn to the table section that gives
    information for your assumed lati-
    tude.

3. Within the section of the assumed
    latitude, turn to the page that gives
    figures that relate to the planet's dec-
    lination in whole degrees.

4. Within the assumed latitude section
    for the planet's declination in de-
    grees, select the subsection that cor-
    responds to the *same* or *contrary*
    name of the declination, depending
    on the hemispheres of you and the
    planet.

K. Having located the proper page in H.O. 249, Vol. II, which contains the assumed latitude, the correct declination in whole degrees, and the appropriate name of declination, you can obtain the final pieces of information.

1. Down the extreme left and right sides of the page are columns entitled "LHA." Proceed down these columns until you locate the one for your LHA (which came from I above). At the intersection of the appropriate LHA and the column of the declination in whole degrees is a series of numbers. The first set has four places and at the top of the column is labeled "Hc" (computed altitude) in degrees and minutes. The second set has one or two places and is labeled "d" and is in minutes. The third set has two or three places and is labeled "Z" and is in whole degrees.

2. Note carefully the degrees and minutes under "Hc." This is your computed altitude, assuming that the planet's declination is in whole degrees, which it almost never is.

Therefore, you must interpolate for that portion of declination which exceeded the whole degrees of declination. The number under "d" is the key to the interpolation to get a completed Hc.

3. Note the plus (+) or minus (−) sign at, or just above, the numbers under "d" as well as the numbers themselves. Also note the remaining minutes of declination not accounted for. Staying within Vol. II, turn to page 242, Table 5, "Correction to Tabulated Altitude for Minutes of Declination." The row across the top ranges from 1 to 60 and is used for "d." Locate the line of numbers for "d." At the edges of the table are two identical columns running from 0 to 59 and headed with the sign for minutes ('). Proceed down either column to the number of minutes of declination that need to be interpolated. At the intersection of the two ("d" and minutes) is the value in minutes for the correction to Hc. (In truth, accidental reverse of "d" and minutes will make no difference, as the an-

swer is the same.) This number of minutes should be added to, or subtracted from, Hc depending on the plus or minus sign previously noted for "d." The sum of Hc in whole degrees and minutes for "d" from Table 5 is the complete Hc for your sight.

4. At this point almost the only remaining step prior to plotting the LOP is to determine the direction from the AP. This is accomplished with the value for Z in degrees noted above. For all northern latitudes in the top left-hand corner of the pages of Vol. II is the legend, "LHA greater than 180°, Zn = Z; LHA less than 180°, Zn = 360 − Z." Therefore, if your LHA is greater than 180°, do nothing. If your LHA is less than 180°, subtract the value of Z from 360°. In both cases the result is Zn or azimuth. Instructions for sailing in southern latitudes are in the bottom left-hand corner.

5. Three critical items of information have been developed in this section. The first two result from a comparison of Hc and Ho. One comparison

provides the direction toward or away from the planet you will place the LOP in relation to the AP. The other comparison provides the distance the LOP will be from the AP. The third item of information is the Zn, which is the exact direction in whole degrees from the AP toward which the planet is located.

L.   The final phase of this process is determining the direction, distance, and placement of the LOP in relation to the AP. This is accomplished as follows:

   1.   Compare the Ho and the Hc. If the Ho is larger than the Hc, then the LOP will be placed toward the planet. If the Ho is less than the Hc, then the LOP will be placed away from the planet.

   2.   The LOP is a line drawn perpendicular to the azimuth line. The distance along the azimuth is the intercept (Int). The Int is the difference between the Ho and the Hc. This difference in minutes of arc corresponds directly to the distance in nautical

miles that the LOP will be drawn across the azimuth from the AP.

3. As indicated, the azimuth (Zn) is the direction in degrees toward which the planet lies, and along which you draw a line through your AP.

4. With pencil and plotter, draw a line through the AP along the Zn, extending away from the planet if the Hc is larger than the Ho, and toward the planet if the Ho is greater than the Hc. Then take the dividers and measure the distance along the longitude scale of the chart or plotting paper of the Int, which was the difference in minutes between the Hc and Ho. Marking from the AP, draw a line perpendicular to the azimuth at the point of intercept. This perpendicular line is the LOP. Your vessel should be located approximately along this LOP. The intersection of multiple LOPs from other bodies constitutes a fix.

M. Compare your LOP or fix with your DR

position. Unless you have made a mistake in the calculation of either the LOP or the DR position, the two should be reasonably close, considering the circumstances. An examination of the steps should resolve significant differences.

8.

STAR LINE
OF POSITION

Star Line of Position

Items needed: Sextant, *The Nautical Almanac* (N.A.), *Sight Reduction Tables* (H.O. 249, Vols. I and II or III), accurate time, plotting paper, graph paper, straightedge, scratch paper, pencil, dividers, and plastic plotter.

A. Note body (star), date, deduced latitude and longitude (DR position). Determine height of eye above water.

B. Depending on skill, experience, and conditions, take a number of sextant shots, recording exact watch time of each.

C. To select the desired sextant reading and

time, proceed as follows: On graph paper, place on the left side a vertical scale in degrees and minutes of arc for sextant readings. Across the bottom, place a horizontal scale for time. Arrange, or plot, sextant readings as recorded for degrees and time. With a straightedge, place a line through the apparent *flow* of the plotted points. Select the raw sextant reading and watch time nearest the drawn line. These are the shot and time to be used.

D. It is necessary to make certain corrections and adjustments to the raw sextant reading selected in C above, as follows:

1. Take raw sextant reading (Hs).   Hs

2. Adjust for sextant index error.   + / −IC

3. The result is corrected Hs.   Corr Hs

4. Subtract dip for height of eye from inside front cover of the N.A.   −dip

5. The result is apparent altitude.

Ha

6. From inside front cover of the N.A., under "Stars and Planets," select the correction for refraction at, or immediately above, the apparent altitude listed. This is the altitude correction factor and is always subtracted.

—Alt. Corrn

7. The result is the observed altitude of the star.

Ho

E. Calculate Greenwich mean time (GMT) from the time of the selected sextant shot as follows:

1. Take raw watch time (WT) in hours, minutes, and seconds.

2. Correct for watch error (WE): subtract for fast, add for slow.

3. Confirm A.M. versus P.M. If later than 1:00 P.M., add 12 hours.

4. The sum of (1), (2), and (3) = ZT (zone time).

5. Add zone description (ZD) to ZT to get GMT. For example, Miami, Florida, is in Zone 5, as zones are 15° wide from Greenwich (80° divided by 15° = 5+ hours); however, during summer, if daylight saving time (DST) is in effect, add one less hour (i.e., 5 − 1 = 4 hours) to ZT to get GMT.

6. If GMT as calculated is greater than 24 hours, subtract 24 hours from time and add one day to get new Greenwich date; the remainder of the GMT becomes the new time.

F. Calculate the Greenwich hour angle (GHA) of Aries as follows:

1. Turn to the proper Greenwich date in the white pages of the N.A. Under the left-hand GMT column, entitled

"Aries," locate the GHA of Aries for the whole hour of GMT. Write down the degrees, minutes, and tenths of minutes shown.

2. To interpolate the GHA of Aries for the remaining minutes and seconds of GMT, turn to the yellow pages entitled "Increments and Corrections" in the back of the N.A. These pages are marked in minutes of time with seconds of time running down the left-hand columns. Two minutes, with 60 seconds each, are displayed on a page. Having selected the correct minute and second of time, take from the "Aries" column the degrees, minutes, and tenths of minutes of arc to add to those of the whole hour of GHA of Aries.

3. The addition of (1) and (2) gives the GHA of Aries in degrees, minutes, and tenths of minutes of arc for the exact hour, minute, and second of time of the sextant shot. In west longitude this corresponds to the geographic longitude of the celestial line of Aries. In east longitude, the GHA of Aries is subtracted from 360° to obtain the celestial line of Aries.

G. Select the sidereal hour angle (SHA) of the star and calculate the GHA of the star as follows:

1. Return to the white pages of the N.A. to the Greenwich date being used. In the center section by the fold is a column entitled "Stars" with the principal ones used in navigation. The list serves for all three days shown on the page.

2. From beside the star being used, note the SHA given in degrees, minutes, and tenths of minutes of arc.

3. Though it will not be used for a while (J below), now is a convenient time to note the Dec (declination) of your star, which is listed right beside the SHA. The Dec is given in degrees, minutes, and tenths of minutes of arc. As the star may be north or south of the equator, the letter N or S indicates declination. This Dec corresponds to the geographic latitude of the star. Hold this data for easy reference.

4. To obtain the GHA of the star, add

the SHA of the star to the GHA of Aries (from F above). If the total exceeds 360°, which it often does, simply subtract 360° and work confidently with the remainder. There are enough circles in celestial navigation without going around twice. The GHA of the star relates to its geographical position. In west longitude this corresponds directly. In east longitude, the GHA of the star is subtracted from 360° to obtain its longitude.

5. The determination of the GHA and the Dec of the star provides the geographic position (GP) of the star. It is from this location that your line of position (LOP) is calculated.

H. For purposes of computation and to select the appropriate tables, it is necessary to assume a position. This is not your DR position; it is merely one nearby that facilitates the use of tables and avoids complex calculations. The initial steps are as follows:

1. In west longitude select the latitude in whole degrees that is nearest your

DR latitude. This is your assumed latitude.

2. Select the nearest longitude that would end with exactly as many minutes and tenths of minutes as the GHA of the star when you took the sight (G.4 above). When in east longitude, change the minutes and tenths of minutes of your DR longitude to equal the difference between the GHA of the star and a whole degree so that, when adding, the assumed longitude comes out to a whole degree.

3. Plotting (1) and (2) on plotting paper gives your assumed position (AP).

I. From the GHA calculated in G and with the assumed longitude determined in H.2 above, it is possible to construct the local hour angle (LHA) of the star, which is the angle measured in a westward direction only from your AP to the meridian of the star. It is essential to accept that when your vessel is in west longitude, LHA is less than GHA because the star must pass Greenwich before it passes you. Conversely, when your vessel is in east longitude, LHA is greater than GHA since the star passes you before it reaches Greenwich. The LHA is calculated as follows:

1. When in west longitude, if your assumed longitude is less than the GHA, the LHA is obtained by subtracting the assumed longitude from the GHA of the star. However, if your assumed longitude is greater than the GHA, the LHA is obtained by *first* adding 360° to the GHA and *then* subtracting your assumed longitude.

2. When in east longitude, the LHA is obtained by simply adding the assumed longitude to the GHA of the star. If the total of your assumed longitude and the GHA exceeds 360°, the LHA is then obtained by subtracting 360°.

3. It should be noted that the assumed longitude was determined to be a meridian of longitude that caused the resulting LHA to be in whole degrees.

J. At this point it becomes necessary to select

between Vol. I (known as the "red" book, because of its red-colored binding) and Vol. II (known as the "white" book) or Vol. III (known as the "blue" book) of H.O. 249 *Sight Reduction Tables.* To facilitate this decision, refer to the Dec of the star as noted in G.3 above. If the declination of the star, either N or S, is greater than 29°, it is essential to use the red book, Vol. I. Use of this book will be described momentarily. If the declination of your star is less than 29°, either N or S, it may be possible to use either book. Brief practice with Vol. I will demonstrate that it is convenient, time-saving, and requires fewer calculations than Vol. II, the white book. Therefore, when the declination is less than 29°, N or S, selection of the volume is a matter of personal preference.

One of the principal benefits of Vol. I is that you can complete much of your work in advance, because it helps identify the stars to work with and gives their computed altitudes and azimuths. With some practice and by essentially working the problem backward you can determine exactly when to shoot the star with the sextant. This time will be at evening or dawn as listed on the right-hand side of the page in the N.A. for the Greenwich date of the sextant shot. A comparison of the sextant's observed altitude with the predeter-mined Vol. I altitude produces your intercept for the line of position. Doing the same thing with multiple stars will let you plot several lines of position and obtain a fix in just min-utes. It may be worthwhile to note that the stars in Vol. I that are printed in capital letters are of the first magnitude and therefore the brightest. Those with little diamonds before them will give the best intersecting lines of position and, therefore, the finest fix. Last, but not least, Vol. I serves as an accurate star finder because you only have to preset your sextant to the stated computed altitude (Hc), adjust for sextant corrections, dip, and refrac-tion by reversing, and point it in the direction given as the azimuth (Zn).

K. To use the red book, Vol. I, proceed as follows:

1. Turn to the reduction table corre-sponding to your assumed latitude. Note that the tables are constructed in whole degrees from Lat 89° N to Lat 0° and to Lat 89°S.

2. On two facing pages for your as-sumed latitude are four sections ar-

ranged by the LHA of Aries at 0° in the upper left-hand corner and ending with 359° at the lower right. Proceed through the table until you locate the LHA of Aries from I above. To the right of the LHA of Aries are the names of seven stars. Your star should be among the seven listed.

3. At the top of the column with your star are two headings, "Hc" and "Zn." Under "Hc" is the computed altitude of the star in degrees and minutes for your assumed position. Also shown is the azimuth (Zn) of the star in whole degrees. This is the direction to the star from your assumed position.

4. As the data provided in Vol. I are not permanent, new volumes are issued for epochs of five years. In each issue, and in Table 5, "Precession and Nutation Correction of Vol. I," are two easy adjustments that may be made to your assumed position (AP) fix, or line of position. These adjustments are called "precession" and "nutation" and relate to the wobble

and turning of the earth. Usually these adjustments improve your accuracy only a mile or two and therefore could perhaps simply be ignored in open oceans. An excellent explanation of the use of Table 5 is provided at the bottom and states that either the line of position or the fix should be moved a short distance in a stated direction. However, most navigators who bother with this adjustment either move the assumed position (AP) and then lay off the LOP from there or wait until they complete their fix and then move it the given distance and direction. Few bother to draw an initial LOP and then apply the correcting factors followed by a corrected line of position.

L. To use the white book, Vol. II (picking up at the end of I above), write the Dec of the star from G.3, the assumed latitude from H., and the LHA of the star from I. It is now possible to enter the tables of H.O. 249, Vol. II, and pursue its method of determining the required computed altitude (Hc) and azimuth (Zn). As indicated, to use this volume the star must be among the 57 shown in the N.A. and must have a declination of 29° or less, either north or south. Note: For latitudes of 40° or more, H.O. 249, Vol. III, is comparably arranged and similarly used.

1. In the tables for latitudes from 0° to 39°, for each whole degree of latitude there are four headings: "Declination (0°-14°) *Same* Name As Latitude," "Declination (0°-14°) *Contrary* Name to Latitude," and two similar headings for declinations from 15° to 29°. *"Same* Name" or *"Contrary* Name" is determined by whether the north or south declination of the star puts it in the same hemisphere as your vessel.

2. Turn to the table section that gives information for your assumed latitude.

3. Within the section of the assumed latitude, turn to the page that presents figures that relate to the star's declination in whole degrees.

4. Within the assumed latitude section for the star's declination in degrees, select the subsection that corresponds to the *same* or *contrary* name

of the declination, depending on the hemispheres of you and the star.

M. Having located the proper page in H.O. 249, Vol. II, which contains the assumed latitude, the correct declination in whole degrees, and the appropriate name of declination, you can obtain the final pieces of information.

1. Down the extreme left and right sides of the page are columns entitled "LHA." Proceed down these columns until you locate the one for your LHA (which came from I above). At the intersection of the appropriate LHA and the column of declination in whole degrees is a series of numbers. The first set has four places and at the top of the column is labeled "Hc" (computed altitude) in degrees and minutes. The second set has one or two places and is labeled "d" and is in minutes. The third set has two or three places and is labeled "Z" and is in whole degrees.

2. Note carefully the degrees and minutes under "Hc." This is your computed altitude, assuming that the

star's declination is in whole degrees, which it almost never is. Therefore, you must interpolate for that portion of declination which exceeded the whole degrees of declination. The number under "d" is the key to the interpolation to get a completed Hc.

3. Note the plus (+) or minus (−) sign at, or just above, the numbers under "d" as well as the numbers themselves. Also note the remaining minutes of declination not accounted for. Staying within Vol. II, turn to page 242, Table 5, "Correction to Tabulated Altitude for Minutes of Declination." The row across the top ranges from 1 to 60 and is used for "d." Locate the line of numbers for "d." At the edges of the table are two identical columns running from 0 to 59 and headed with the sign of minutes ('). Proceed down either column to the number of minutes of declination that need to be interpolated. At the intersection of the two ("d" and minutes) is the value in minutes for the correction to Hc. (In truth, accidental reversal of "d" and minutes will make no difference, as the an-

swer is the same.) This number of minutes should be added to, or subtracted from, Hc depending on the plus or minus sign previously noted for "d." The sum of Hc in whole degrees and minutes for "d" from Table 5 is the complete Hc for your sight.

4. At this point almost the only remaining step prior to plotting the LOP is to determine the direction from the AP. This is accomplished with the value for Z in degrees noted above. For all northern latitudes in the top left-hand corner of the pages of Vol. II is the legend, "LHA greater than 180°, Zn = Z. LHA less than 180°, Zn = 360 − Z." Therefore, if your LHA is greater than 180°, do nothing. If your LHA is less than 180°, subtract the value of Z from 360°. In both cases the result is Zn, or azimuth. Instructions for sailing in southern latitudes are in the bottom left-hand corner.

N. At this juncture the information from both Vol. I and Vol. II has been obtained and the data can be handled in essentially the same manner. Three critical items of information have been developed. The first two result in a comparison of Hc and Ho, from D above. One comparison provides the direction toward or away from the star you will place the LOP in relation to the AP. The other comparison provides the distance the LOP will be from the AP. The third is the Zn, which is the exact direction in whole degrees from the AP toward which the star is located. The final phase of this process is determining the direction, distance, and placement of the LOP in relation to the AP. This is accomplished as follows:

1. Compare the Ho and the Hc. If the Ho is larger than the Hc, then the LOP will be placed toward the star. If the Ho is less than the Hc, then the LOP will be placed away from the star.

2. The LOP is a line drawn perpendicular to the azimuth line. The distance along the azimuth is the intercept (Int). The Int is the difference between the Ho and the Hc. This difference in minutes of arc corresponds directly to the distance in nautical miles that the LOP will be drawn across the azimuth from the AP.

3. As indicated, the azimuth (Zn) is the

direction in degrees toward which the star lies, and along which you draw a line through your AP.

4. With pencil and plotter draw a line through the AP along the Zn, extending away from the star if the Hc is larger than the Ho, and toward the star if the Ho is greater than the Hc. Then take the dividers and measure the distance along the longitude scale of the chart or plotting paper of the Int, which was the difference in minutes between the Hc and Ho. Marking from the AP, draw a line perpendicular to the azimuth at the point of intercept. This perpendicular line is your LOP. Your vessel should be located approximately along this LOP. The intersection of multiple LOPs from other bodies constitutes a fix.

O. Compare your LOP or fix with your DR position. Unless you have made a mistake in the calculation of either the LOP or the DR position, the two should be reasonably close, considering the circumstances. An examination of the steps should resolve significant differences.

9.

NAVIGATION WITHOUT SEXTANT

Navigation without Sextant

Items needed: *The Nautical Almanac* (N.A.), *Sight Reduction Tables* (H.O. 249, Vol. II or III), accurate time, plotting paper, graph paper, scratch paper, pencil, dividers, and plastic plotter.

A. Note body (sun) (while the concept applies to the moon, stars, and planets as well, this examination will be of the body most commonly used; when applying this procedure to the moon, use the "Altitude Correction Tables—Moon," found on the inside back cover of the N.A.), date, deduced latitude and longitude (DR position). Select upper or lower portion (limb) of the sun for a one-time effort.

(To increase your chances of success with the sun, there is the possibility of taking both upper and lower limb readings. This is fine, but keep the mathematical calculations separate, as the main correction for the semidiameter of the sun is added for the lower limb and subtracted for the upper limb.) Determine height of eye above water.

B. Try to judge the instant at which the selected limb crosses the horizon (when it appears in the case of a rising sun, or disappears when setting) and note the exact watch time. In the case of the moon, to some degree, and the stars and planets, in particular, it is recommended that you use binoculars to increase light-gathering ability and to improve the estimate of when the body crossed the horizon. Use of this technique for stars and planets is generally limited to clear nights with substantial moonlight providing a clearly defined horizon. At dawn and dusk, when these bodies become more visible than during the day, the ambient light usually impairs your ability to pick up the bodies as they rise or be certain of their departure when they set. There is, however, the possibility of "bobbing" a limb of the moon or the planets and stars at any time during the night. The technique may be used for either a selected body disappearing over the horizon to the west or one scheduled to appear in the east. For a disappearing body, a low position in the boat is desirable; for an appearing body, position yourself as high as possible. Either way, by standing upright and stooping, it is possible for a few moments to make the body appear and disappear. As this occurs, note the time and judge your average height of eye to obtain the "dip" factor.

C. Since you have no series of readings, you cannot plot them on graph paper. Consequently, you cannot be entirely confident that the timing was correct and was not overly influenced by the motion of the vessel. Uncertainty of 20 seconds as to when the limb or the body actually appeared when rising, or disappeared when setting, produces an uncertainty of approximately five nautical miles in your calculated line of position (LOP). However, if you are without a sextant, accuracy of plus or minus five miles might well be considered a magnificent success.

D. Even though no sextant was actually used, the calculations for producing the observed altitude (Ho) are the same. The difference is that the sextant altitude is zero, since the observation was made when the limb or body was, in effect, brought down to the horizon. Obviously, there is no sextant error correction. Therefore, the calculation is worked as follows:

1. Take observed reading (0°00.0′).   0°00.0′

2. Subtract dip for height of eye found inside front cover of the N.A.   —dip

3. The result is apparent altitude.   Ha

4. From page A3 of the N.A., under "Sun," select the month period and upper or lower limb correction.   Alt. Corrn

5. Due to the low Ha, it is desirable to correct for the effects on the atmosphere of barometric pressure and temperature.   See

page A4, "Altitude Correction Tables—Additional Corrections," in the N.A. Lacking exact barometric and temperature information, it is worthwhile making a guess in order to obtain the best adjustment possible.

+ / —Add'l
Corrn

6. The result is the observed altitude of the sun.

Ho

E. From this point on, the calculation of GMT, GHA, LHA, assumed latitude and longitude, and the use of the N.A. and H.O. 249, Vol. II and III, are exactly the same as for the sun in Chapter 4, "Sun Line of Position," beginning with section E. For other celestial bodies, refer to the applicable chapter.

F. One additional point must be made, however, about the determination and construction of the intercept for the LOP. Due to refraction, the limb of the sun (or other body) is always below the actual horizon. The effect of atmosphere is to permit observation when otherwise the limb would have disappeared. As a result, there may be some positive and negative numbers in your calculations. In such instances, the corrections must be determined algebraically. For negative altitudes, a negative correction is numerically added, i.e.:

$$Ho = -0°30.1'$$
$$\underline{Hc = -0°26.4'}$$
$$Int = -0°03.7' \text{ (away)}$$

Normally, with the Ho being greater than Hc, the intercept would be *toward* the sun (or other body). Since the intercept is negative, however, the LOP is placed *away* from the sun. Also algebraically, a positive correction must be numerically subtracted, i.e.:

$$Hc = +0° \ 25.7'$$
$$\underline{Ho = -0° \ 19.2'}$$
$$Int = -0° \ 44.9' \text{ (away)}$$

In both instances, you can appreciate that the rule "Ho more than Hc, place LOP *toward* the body" still applies. It just takes some reassurance that a negative Ho with a larger number than the negative Hc is actually just "more negative" and therefore smaller, or that a negative Ho must pass all the way up through zero to be compared with a positive Hc. The result is addition. Again, since a negative Ho is obviously smaller than a positive Hc, the intercept must be *away* from the assumed position (AP), with the intercept being the sum.

G. While this book has gone to great lengths to provide detailed, step-by-step instruction on the mechanics involved in each type of sight reduction problem, even with some fairly substantial repetition among certain chapters, there is no sense repeating where nothing new is added. To complete the calculations here, therefore, note the Ho obtained above, along with the intercept information, take the uncorrected watch time from the sighting, and refer to the chapter for the sun line of position (or the appropriate chapter for some other body) at the corresponding section and proceed in the normal fashion from that point.

10.

THEORY—BRIEFLY

Theory—Briefly

As stated in the beginning, this book empha-
sizes the "how to" rather than the "why."
Presumably everyone interested in celestial navigation wishes to master the "how to." Since the
"how to" can become rather mechanical, de-
spite its basis in highly complex trigonometric
formulas, it is possible to write a useful and reasonably concise description of the steps involved. The preceding chapters were written for
those wishing such guidance.

The explanation of the "why" of celestial navigation can occupy for a lifetime the minds of
men vastly more able than mine. Therefore, due
purely to intimidation by the potential complex-
ities one could pursue, it would be easy for me

to avoid the "why" side. Also, as the interest levels and capacities of readers may vary, I am uncertain of the degree (no pun intended) to which I should explore the technical facets of celestial movement and spherical triangles. Nevertheless, this chapter provides a very brief outline of the basic factors of the functions involved.

First, I urge the reader to acquire the Bowditch book. Its size is awesome, but the style is intelligible. Furthermore, it is divided into many diverse sections, thus permitting study of reasonable quantities of material in succession. Most important, because it is a government publication, it happens to be very inexpensive for the quantity and quality of information it contains.

In addition, many popular books on the market offer a combination of theory and practical guidance. Their emphasis and perspectives differ. What the reader may find impossible to understand in one book may seem crystal clear in another. From my experience and the comments of others, I believe that three other books will be complementary and help provide a good "fix" on both theory and practice. The three books are: *Celestial Navigation for Yachtsmen* by Mary Blewitt, *Self-Taught Navigation* by Robert A. Kittredge, and *Common Sense Navi-*

gation by Hewitt Schlereth. All three are current and applicable to *The Nautical Almanac* and H.O. 249, *Sight Reduction Tables for Air Navigation.* So, let us continue.

As you know, the sextant is an instrument for measuring angles between a celestial body (sun, moon, star, or planet), the navigator's eye, and the horizon of the observer on the surface of the earth.

Due to geometric and atmospheric conditions, what you believe you see varies somewhat from what actually is. The use of four "givens" (the North and South Poles, the Greenwich meridian, the equator, and the exact time) in concert with almanacs and sight reduction tables permits you to determine the various interrelationships. The easy use of the almanac makes it possible to determine for an exact second of time where a celestial body is relative to a prime meridian (Greenwich) and the equator. Thus, while the celestial body is far out in space, its relative geographic position, or latitude and longitude, are known.

The earth is 360° in circumference, and each degree, in turn, is divided into 60 minutes. Each such minute of arc equals 1 nautical mile. Therefore, it follows that 360° multiplied by 60 minutes equals 21,600 nautical miles, roughly the distance around the earth when measured

FIGURE X.1

GLOBE, BASIC REFERENCE
POINTS AND SUN PASSING
GREENWICH MERIDIAN LOCAL NOON

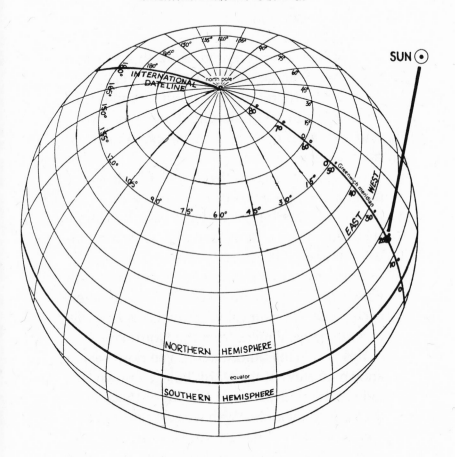

through the North and South Poles. Given the location, or geographic position (GP), of the body above earth, it is a simple calculation to determine its distance from the North Pole. See Figure 10.1.

Directions are presented as north and south from the equator and east and west from Greenwich. Therefore, 90° north or south is as far as you can go in the direction of the poles. And 180° east or west is as far as you can go from Greenwich.

Therefore, it follows that if the latitude of the celestial body is known, its distance from the North Pole can be calculated easily. By simple subtraction, if the body has a GP with a northerly latitude of 18°, it is 72° (90° − 18° = 72°) from the North Pole. If it is 72° from the North Pole, and each degree contains 60 minutes, each equal to a nautical mile, the GP is 72 × 60, or 4,320 nautical miles from the North Pole.

This is valuable information. We now have a known location of the North Pole, a known latitude of the body, and the known distance from the body to the pole. We have one side and two corner points of a triangle. And we can do more.

Since the GP of the body was provided through the almanac at a time measured from Greenwich mean time (GMT), we are able to measure the longitude of the body from the

FIGURE X 2

GREENWICH HOUR ANGLE
DECLINATION AND
GEOGRAPHIC POSITION OF SUN

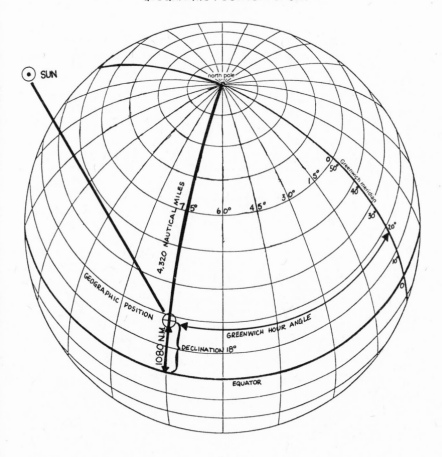

FIGURE X.3

SPHERICAL TRIANGLE

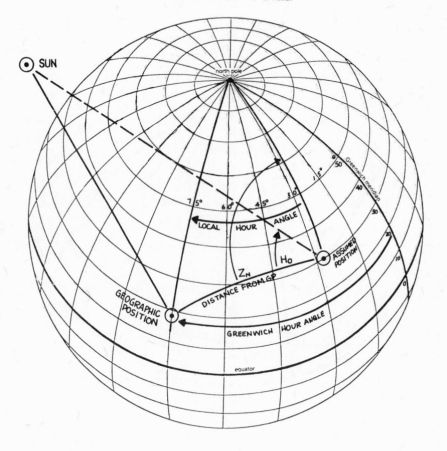

Greenwich meridian. Thus we have obtained the angle of the meridian of the body from the Greenwich meridian at the intersection of these two meridians at the North Pole (or from the South Pole, if the body is in the Southern Hemisphere). See Figure 10.2.

This angle is very useful. It is known as the Greenwich hour angle (GHA) because it is the angle formed at a specific moment of GMT relative to the Greenwich meridian.

While the GP of the body has been determined by the use of the almanac for the moment of the sextant shot, the sextant angle obtained by the shot is useful in determining the distance of the navigator from the GP of the body. As spheres are spheres from any direction, so it follows that the sextant angle corrected for index error, dip, atmospheric refraction, and so on (i.e., the Ho) can also be subtracted from 90° to determine the distance of the navigator from the GP of the body. Thus, an Ho of 58° would place the observer 1,920 nautical miles from the GP of the body (90° − 58° and 32° × 60 minutes = 1,920 nautical miles).

Now we are making some real progress. We know the exact GP of the body for an exact moment in time and our distance from that GP. See Figure 10.3.

If we had our exact direction from the GP of the body and a sufficiently large chart, we could measure with adequate accuracy our exact location. Unfortunately, we do not have a sufficiently exact measure of the direction toward the GP to avoid the computations of spherical trigonometry. The two best examples for reflection and understanding are (1) the sun as it passes your local meridian and thus eliminates the need for the computations, and (2) the pole star, Polaris, which is sufficiently close to the North Pole to *almost* permit direct measurement.

Fortunately, however, the complex spherical trigonometric calculations have been computed, and their results are contained for easy reference in the sight reduction tables. The use of the almanac and the appropriate reduction tables is, of course, the basis of this book.

11.

PLOTTING PAPER AND LINES OF POSITION

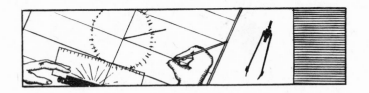

Plotting Paper and Lines of Position

In the preceding chapters, detailed instructions were provided on how to determine the assumed position (AP) of the vessel, the geographic position (GP) of the celestial body, the azimuth (Zn) line, comparison of Ho and Hc to produce the intercept (Int), and placement of the line of position (LOP) relative to the AP. For anyone who is plotting his LOP directly on the navigational chart or who is familiar with the use of plotting paper, that description should prove adequate. However, for those unfamiliar with plotting paper, a brief explanation may help.

First, plotting paper is nothing more than a work sheet that you use instead of messing up your actual charts. Second, you can readily make plotting sheets yourself (as described in Bowditch) or purchase them. Purchased plotting paper either comes with predetermined latitudes or the user must determine the latitude and arrange the longitudes accordingly. The latter is called universal plotting paper since its use is unrestricted. I recommend it because it is inexpensive, "universal," and reusable for a generous number of LOP constructions. Of the several types available, the plotting sheets published by the Defense Mapping Agency appear excellent and particularly easy to use.

This universal plotting paper is prepared for use in just a few simple steps. First mark the middle horizontal line with the latitude of the AP. The other two horizontal lines will be latitudes of one degree difference, either greater or lesser, than the center line. At this point the latitude scale has been arranged.

To make the universal plotting paper for longitude, draw a line perpendicular to the middle latitude line, passing through the number on the arc of the central compass rose that is the same as the number of degrees of latitude. Thus, if the latitude of the AP is 25°, draw a vertical line through the number 25 in the compass rose and perpendicular to the lines of latitude.

In the center of the page is a vertical line with a scale in minutes marked along its length. Place a second longitude line the same distance from the center scale line on the other side. Additional lines of longitude may be marked across the page maintaining this same distance. The center vertical line with the minutes scale is to be labeled with the whole degree closest to that of the whole degree of longitude of the AP. The other meridians of longitude may be marked with their appropriate whole degrees.

The scale in minutes along the center line of longitude is used for plotting longitude and measuring distances in nautical miles. In the lower right-hand corner of the plotting paper is a longitude scale. This may be used by selecting the line of longitude shown closest to your chosen center horizontal line of longitude and marking the corresponding points of longitude on the plotting sheet as desired.

The universal plotting paper is now complete for use as described in the preceding chapters. With a little practice, you will come to appreciate its usefulness. You can keep your charts uncluttered and obtain more precise navigational fixes.

Figure 11.1 shows the plotting of an LOP on universal plotting paper of the type described. Note that the AP is located on a whole degree of latitude. In this case the LOP was toward the celestial body, and the intercept (Int) was measured along the azimuth line in nautical miles corresponding to a difference between the Ho and the Hc. The intersection of multiple LOPs constitutes a fix.

At this point it is tempting to launch into a comparison of deduced position, estimated position, most probable position, force vectors, winds, currents, and set and drift. However, such factors are common to *all* navigation and pilotage, not just celestial, and therefore must be defined as outside the limited scope of this book. Furthermore, these considerations and their interrelationships have been well described in other texts and I have nothing new to add. Just remember their importance for successful navigation.

It is appropriate, however, to include information on advancing or retarding lines of position. While common to all navigation and pilotage, the ability to advance or retard lines of position is critical to celestial navigation and the determination of a running fix. A fix formed by the intersection of multiple LOPs from sights made at about the same time may give the best indication of the vessel's location, but it is often impossible to obtain one. Morning and evening twilight provide the best opportunities for multiple body sights. Certain days may offer a good shot at both the sun and the moon.

81°

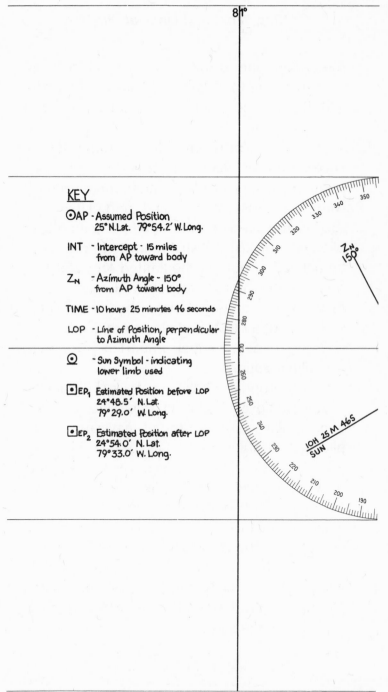

KEY

⊙AP - Assumed Position
    25° N. Lat. 79° 54.2′ W. Long.

INT - Intercept · 15 miles
    from AP toward body

Z_N - Azimuth Angle - 150°
    from AP toward body

TIME - 10 hours 25 minutes 46 seconds

LOP - Line of Position, perpendicular
    to Azimuth Angle

⊙   - Sun Symbol - indicating
    lower limb used

⊡EP₁ Estimated Position before LOP
    24° 48.5′ N. Lat.
    79° 29.0′ W. Long.

⊡EP₂ Estimated Position after LOP
    24° 54.0′ N. Lat.
    79° 33.0′ W. Long.

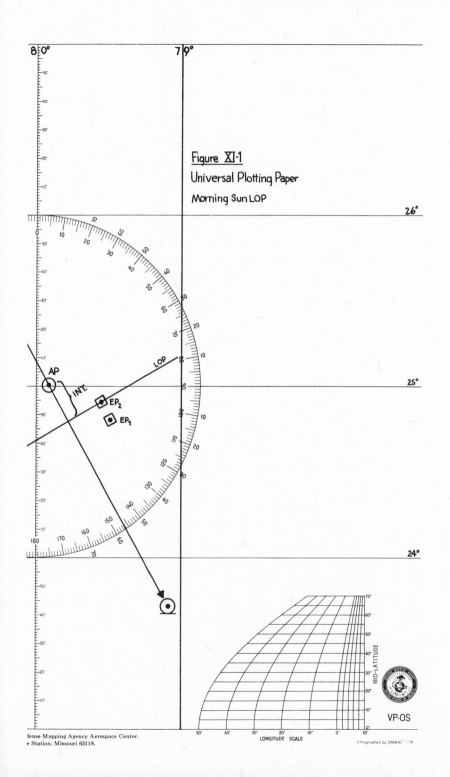

Figure XI-1
Universal Plotting Paper
Morning Sun LOP

LONGITUDE SCALE

MID-LATITUDE

VP-OS

Lithographed by DMAAC

Much of the time, however, the navigator has only the sun to use. But the sun is both the brightest and the fastest-moving object in the sky and thus commonly serves as the principal celestial reference, particularly for navigation well out to sea. For example, if the navigator takes his morning sight at 0900, he knows the vessel should be located approximately along the LOP, but he knows not just where. Since the sun traverses the sky at 15° per hour, in three hours the bearing will have expanded by 45°. In taking his noon sun shot the navigator can obtain an actual fix with both latitude and longitude. But he can add to his confidence by advancing the morning sun line to that of noon.

The process for moving the morning sun line is to return to the plotting sheet and draw a line for the course line and the miles traveled since the time of the LOP. Using parallel rulers, mark off a line parallel to the old LOP across the course line at the point that allows for the distance traveled. The intersection of this advanced line with the noon line should correspond approximately to the longitude line provided by the noon sight. If a noon fix has been obtained with longitude determined from GMT, you can simultaneously provide a running fix and confirm the noon calculations.

At about 1500 hours, or 3:00 P.M., a new sun

shot can be taken and a fresh LOP drawn. The navigator then can either advance the noon fix and remain with an advanced-line running fix, or retard the new LOP back to noon by marking off the course line and miles traveled since noon and drawing a line parallel to the latest LOP back toward the earlier position. The intersection of the morning and afternoon LOPs, separated as they are by 90°, should give an especially good confirmation of the noon fix. To the extent that there is a serious inconsistency in the anticipated confirmation, the navigator should consider the set and drift of current, unusual leeway or wind conditions, or the accuracy of sights, reductions, and speed and course calculations.

While the preceding example was concerned with the sun, the practice can be applied to all combinations of celestial bodies. Indeed, the experienced navigator can create some very ingenious combinations of pieces of information to help understand his position and the effects of the environment in which he is operating.

When trying to establish the vessel's location with a fix, the navigator must bear several considerations and alternatives in mind. First, a reasonable spread must be available between the celestial bodies or other objects being used. Just as it serves little purpose to calculate two LOPs from objects right next to each other, there is little value in LOPs from objects 180° across the horizon. If only two LOPs are available, it is best if they are approximately 90° apart, although as little as 40° will generally produce satisfactory results. If three bearings are available, spreads of approximately 120° should produce excellent results.

Fixes, of course, can be established from objects other than multiple celestial bodies. Bearings or distances from shore or objects whose height or position is known can also be used.

The use of an LOP from a single celestial body in conjunction with soundings and a chart showing depth can, under certain circumstances, be adequate to establish your approximate location. Certainly when approaching shore or shoal waters, it is wise to help confirm or establish one's location by taking soundings. It is an underappreciated fact that in many circumstances knowing where one *is not* can be of just as much value as knowing where one is.

12.
OTHER INFORMATION OF INTEREST

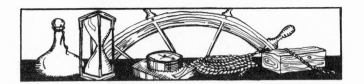

Other Information

A.   One aspect of continual confusion for most navigators who do not have a background in mathematics, engineering, or physics is the conversion of a circle to its parts and those parts to time. As you might expect, *The Nautical Almanac* has a table entitled "Conversion of Arc to Time." It is located at the beginning of the yellow pages and is worthy of study.

To complement the table, I personally find an expression of arc in nautical miles to be useful. Also, since time zones and the speed of rotation of the earth are measured in time, arc, and nautical miles, a graphic expression of the interrelationships helps me keep clearly in mind whether I am dealing with minutes of

arc or of time and approximately how far or
how fast that may be. See Figure 12.1.

Figure 12.1

ARC/DISTANCE/TIME OF ROTATION

Arc	Nautical Miles	Time
360°	21,600	24 hr.
270°	16,000	18 hr.
180°	10,800	12 hr.
90°	5,400	6 hr.
60°	3,600	4 hr.
30°	1,800	2 hr.
15°	900	1 hr.
10°	600	40 min.
7½°	450	30 min.
5°	300	20 min.
1°	60	4 min.
60′	60	4 min.
45′	45	3 min.
30′	30	2 min.
15′	15	60 sec.
10′	10	40 sec.
5′	5	20 sec.
1′	1	4 sec.

Remember, time zones are 15° wide (360°
divided by 24 hours = 15°), and the sun is

scheduled to have its local apparent noon in the center of each time zone, or at the 7½° point.

B. As accurate time is a principal, in fact critical, factor in celestial navigation, it is essential to have, or have access to, this exact piece of information.

A good ship's clock, either spring wound or with a quartz timing mechanism, should provide an adequately accurate measurement of time for navigation. A stopwatch or an ordinary wristwatch, when used in conjunction with the ship's clock, can permit you to take accurate time on deck for a round of shots. Most vessels maintain or confirm time by receiving radio time signals. In the absence of a radio, however, I would prefer and recommend that an excellent watch of chronometer quality be carried. Although I favor a quartz, many navigators are turning to water-resistant LCD watches that have stopwatch, dual time zones, and other features.

By international agreement, time signals are broadcast on five radio bands for dedicated standard frequencies. Most maritime nations broadcast time signals several times daily. Information on worldwide time signal systems is available in Publications No. 117A and

117B, *Radio Navigational Aids,* and volume 5 of the *Admiralty List of Radio Signals.*

Of great interest to most navigators are certain radio time signals provided by the United States. Stations WWV, in Colorado, and WWVH, in Hawaii, emit signals on multiple frequencies to enhance prospects that at least one will be received anywhere in the world. Propagation and reception characteristics obviously affect such signals, and certain ones are occasionally heard better than others. Nighttime is often considered favorable for radio reception. Both WWV and WWVH broadcast time ticks signaling each second on 2.5, 5.0, 10.0, and 15.0 MHz. WWV also broadcasts on 20.0 MHz. Voice announcements (male voice for WWV and female voice for WWVH) are made for each minute. It is especially helpful that the hour reference is to GMT. Thus the navigator can confirm his GMT without regard to time zones.

In addition, both WWV and WWVH provide storm information. WWV provides this for the Atlantic region on the eighth, ninth, and tenth minutes after the hour. WWVH gives storm information for the Pacific 12, 11, and 10 minutes before the hour.

A device of increasing popularity among student and practicing navigators is a radio time cube available through Radio Shack. This is a small, battery-operated, inexpensive, and highly serviceable portable radio that receives on three (5.0, 10.0, and 15.0 MHz) of the frequencies used by WWV and WWVH.

It also is important to have the day and date available. It is surprisingly simple to become completely confused as to the date even when you have been at sea for only a few days. This can be a serious problem, obviously.

To help avoid confusion, draw a line across, or X out, each day in the almanac when you are finished with it. However, do not tear it out and throw it away. You may have to go back and reconstruct some calculations, particularly if you have been sick, if you have experienced severe weather, or if you have been working while extremely fatigued.

C. As the use of computers and calculators has spread into nearly every corner of our lives, it must be appropriate that they come aboard our little vessels. In truth I have nothing against them. Calculators or pocket-sized minicomputers can greatly facilitate the calculations of sight reduction. However, there were no doubt old salts who thought precomputed sight reduction tables an abomination that should not be used in favor of mathematically solving the spherical triangles by hand. In a sense they were right; without the tables the next generation of navigators would, with very few exceptions, be absolutely lost.

So it is that many people, myself included, view these flashing digital light things with mixed feelings. While they are certainly not to be feared (since they work so well and so quickly), I am unconvinced that they should be loved either.

You may not know or wonder why officers on naval and merchant ships still take daily sextant readings in this day of electronic and satellite navigation. It is partly tradition; but it is also to keep the historic basic skills in good practice so, if and when necessary, they can still be called upon to produce reliable information.

One day soon, if it has not happened already, some navigator is going to leave his calculator on too long and run down its batteries, drop it and knock it to pieces, or find it has gotten wet or succumbed to prolonged humidity and a hostile environment. At that moment a truth is going to become self-evident. The "navigator" was no navigator at all. To paraphrase the old adage about being up the river without a paddle, he will find himself out to sea without a calculator—and quite literally at that.

Now, having expressed myself with such certainty, let me offer several examples of times and situations where the navigational calculator can be of value.

First, the calculator can be used as a check for your hand-calculated worksheets. While this does not give much benefit in speed, it does keep the basic skills intact and permits rapid confirmation of your efforts.

Second, use of the calculator may be condoned, if not recommended, when you have four, five, or six sightings to reduce. Under normal circumstances you might pick the best two or three and skip the rest. However, after having done the first several in the traditional manner, I can understand that a rapid reduction of the remaining sights might increase the precision of the fix by adding data that otherwise would not have been included.

A third situation favoring calculator computation is where time is a factor, such as in a potentially dangerous situation when immediate information may be critical.

The fourth and most common situation where the calculator may be useful is when you have a crew member aboard who insists on "helping" with the navigation but who is too stupid to learn, or too tiresome to teach, the older, traditional, and more dependable procedures and use of sight reduction tables.

Whatever the circumstances, those who have responsibility for the safe navigation of their vessels need to retain facility with the routines for the reduction of sights for the various celestial bodies. While regular practice is recommended, it is not always possible. Thus it may be that both students and experienced seamen will find good use for this book.

D. Since one's estimated position is very often the most important element in determining or confirming an actual location, anything that can help confirm the estimated position is worthwhile. Obviously, the speed of the vessel through the water is a critical input, even if current and drift are uncertain.

In addition to the use of Walker's Logs, knotmeters, and other devices, there is an ancient technique for measuring speed through the water. It has no parts, as such; it is free; and it can be used for confirmation of the output of other devices or in their stead when they are not functioning—which is often.

Using this method, you convert to speed the length of time it takes for the boat to pass by or leave behind an object tossed in the water. Either of two means will work adequately. The first is to measure a distance along the deck and toss a chip of wood, a cigarette butt, or what-have-you into the water; then measure the time it takes for the object to "arrive" at the other end of the measured distance. The same thing can be accomplished by tying a string of known length to a chip of wood (making a kind of ship's log) and tossing it overboard astern. You then measure the time it takes for the string to pull taut. That is the time the vessel required to pass the specified distance. The accuracy of speed measurement can often be improved by increasing the distance along the deck or the length of the string. While all this seems obvious and fairly simple, the truth is that distance/rate/time problems have tormented man for centuries. That is why special slide rules specifically designed for such relationships are so popular.

For those who may be committed to confirming calculations, the following data are provided: The formulas used are $D = R \times T$; $R = D/T$; $T = D/R$. One nautical mile is 6,076.11549 feet or 1,852 meters. Distance traveled in time (rate) is expressed in knots.

Time is in hours of 60 minutes, each with 60 seconds, or 3,600 seconds per hour.

Many relationships may be calculated from these data. For the purposes of this book, however, two specific distances have been calculated. These distances, when measured in seconds by a vessel passing through water, indicate speed in knots. The distances used are 101.26861 feet (30.87284 meters) and 50.63430 feet (15.3716 meters); they shall be referred to as 101'3" and 50'8". See Figure 12.2. Unfortunately, shorter distances, such as 25 feet, must be measured in fractions of seconds and therefore could require a stopwatch, which may not be available. From the information given, the calculation is easy, however.

As the two distances used will exceed the deck length of most cruising yachts, it will be necessary to use the toss-it-over-the-stern method. Using this technique creates a triangle, the height of which is the distance from the water to the point where the string is tied near the transom. (Note: Tie it. Do not hold it; you will lose it.) Assuming the string is tied to the stern cleat, a height of approximately three feet, more or less, will be common on smaller boats. For the 101'3" distance, the additional length of line necessary (for the hypotenuse of the triangle) is almost unmeasurable. For the shorter distance, the additional length of line is less than an inch and may be ignored.

The construction of the ship's log may be quite simple. Ideally, it should be large enough to bring the line taut with an obvious jerk. Yet it should be small enough to bring aboard easily at high vessel speeds. Beyond that, size is not important. Good results may be obtained with a piece of two-by-four one foot or less in length.

Selection of the string, however, does require some care. It should be stout enough not to break easily or wear out quickly. Also, very small stuff is difficult to handle. Approximately ⅛ inch to ¼ inch seems optimum. Dacron or polypropylene is recommended over either braided nylon or monofilament fishing line because the latter two stretch. You do not want to use a 50-foot line that becomes five feet longer. Personally, I prefer the polypropylene because of its surface texture and because it floats.

An added benefit is that a brightly colored floating line of about ¼ inch diameter securely fastened to a cleat and with the ship's log well tied to the other end may give good service if someone falls overboard, even at night. At six knots the unfortunate soul has a full 10 seconds to get to the 101'3" line. This assumes, of course, that it is thrown to him the instant he hits the water.

It is fair to mention that this technique for determining vessel speed is subject to error and question. While theoretically quite accurate, in practice it is less than perfect. Factors affecting this measurement are the set and drift of any currents and the leeway of the vessel. Variation in vessel speed between times of measurement can also be significant. Inaccuracies of timing and foul running of the line will also produce distortion. In heavy seas, running down the face of a wave or banging into one can increase or decrease the vessel's apparent speed through the water. If you are uncertain of your speed and not comfortable with a single reading of the ship's log, take several measurements, discarding any that appear unreasonable and averaging the rest.

Figure 12.2

SHIP'S LOG TABLE

Speed in Knots	Time in Seconds For 101'3"	Time in Seconds for 50'8"
1	60.0	30.0
2	30.0	15.0
3	20.0	10.0
4	15.0	7.5
5	12.0	6.0
6	10.0	5.0
7	8.6	4.3
8	7.5	3.8
9	6.7	3.3
10	6.0	3.0
11	5.5	2.7
12	5.0	2.5
13	4.6	2.3
14	4.3	2.1
15	4.0	2.0

E. It is unfortunate that many sailors seek to avoid the use of the *Sight Reduction Tables* and depend on noon sights and *The Nautical Almanac* to obtain their position. While the almanac allows one to avoid using the H.O. 249 tables, it requires numerous sights, plotting, graphing, and careful workmanship. An experienced navigator knows that the sun can be blocked from view just when he is trying to sight high noon. While he may be foiled in his efforts to secure his longitude, any sight obtained within several minutes of local apparent noon can be used to work up a functionally adequate latitude. The technique is simply to complete a sun line of position. As the sun was reasonably near its zenith, the LOP will be perpendicular to its azimuth, and thus approximately parallel to lines of latitude. Thus, an almost-line-of-latitude is secured. In bad weather close to shore it may be the best you can do. In good weather, far from land, it may be all you need or wish.

F.   A brief word on using stars is in order. They are easy to find and recognize with very little practice. The last part of the white pages in *The Nautical Almanac* offers good star charts with major constellations drawn in to assist in star location. Also, the N.A. presents information on sunrise, sunset, and morning and evening twilight for numerous latitudes on each right-hand page. The use of these data can help you determine the time period best suited for shooting a round of stars. A clear explanation of the use of the table is provided in the back of the almanac.

Star sights, and resulting LOPs, are excellent. Give a couple of them a try. You will be surprised and pleased. Learn to use Vol. I of the *Sight Reduction Tables;* it is ridiculously easy.

G. Something else that deserves comment are the strips of blank space in the *Sight Reduction Tables,* Vols. II and III. The blank strips correspond to impossible LHAs. In other words, for an object to have an LHA that would appear in the blank spaces, the object would have to be almost exactly on the other side of the world. But since you cannot see it from where you are, there is no purpose in including the data. If your calculations indicate an LHA that falls in the blank strips, check the computation of Greenwich mean time. The chances are excellent you have made an arithmetic error of exactly 12 hours. Furthermore, you were probably using a watch or ship's clock with a 12-hour face and forgot to make the P.M. correction by adding 12 hours to the time shown.

A last point regarding time is to make certain the minute and second hands are properly aligned when setting your timepiece. Confusion as to which minute you are using will produce an automatic error of 15 nautical miles.

H. For those seeking a sample problem to work for various celestial bodies, excellent examples are provided in *The Nautical Almanac* and the *Sight Reduction Tables*. The use of these references and their illustrations as a direct companion piece to this cookbook of celestial navigation has the dual benefit of familiarizing the reader with those sections and permitting the examples to remain current. *The Nautical Almanac* is a wonderful book containing tremendous amounts of truly interesting and useful information. It is a shame that so many navigators merely learn noon sights and never bother to become close friends with it. Just the fact that the print is perhaps a little small should not dissuade you. The format and content are excellent and quite clear considering the subject matter. Finally, *American Practical Navigator,* or "Bowditch," is again recognized as the greatest source of information for study and understanding.

13.

AND IN
CONCLUSION . . .

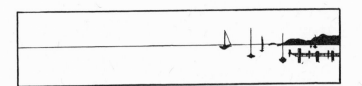

And in Conclusion

Perhaps no adventurer has captured man's spirit and desire to pursue a full and vigorous life better than Jack London. His book, *The Cruise of the Snark* (1911), captivated readers wth experiences and mishaps in the Pacific just after the turn of the century. Setting off in his forty-three-foot *Snark,* and accompanied by his wife, Charmian, his friend Roscoe, and a shipmate named Martin, London had no idea how to navigate at sea. Roscoe, who was to serve as navigator, was also to teach London the necessary skills. However, Roscoe, once having mastered the procedures of sight reduction, demurred and "was like unto a god, and he carried us in the hollow of his hand across the blank spaces on the chart." London determined to teach himself. As he did, he wrote about it. The following excerpt is from *The Cruise of the Snark:*

"And now, in simple language, I shall describe how I taught myself navigation. One whole afternoon I sat in the cockpit, steering with one hand and studying logarithms with the other. Two afternoons, two hours each, I studied the general theory of navigation and the particular process of taking a meridian altitude. Then I took the sextant, worked out the index error, and shot the sun. The figuring from the data of this observation was child's play. In the 'Epitome' and the 'Nautical Almanac' were scores of cunning tables, all worked out by mathematicians and astronomers. It was like using interest tables and lightning-calculator tables such as you all know. The mystery was mystery no longer. I put my finger on the chart and announced that that was where we were. I was right, too, or at least I was as right as Roscoe, who selected a spot a quarter of a mile away from mine. Even he was willing to split the distance with me. I had exploded the mystery; and yet, such was the miracle of it, I was conscious of new power in me, and I felt the thrill and tickle of pride. And when Martin asked me, in the same humble and respectful way I had previously asked Roscoe, as to where we were, it was with exaltation and spiritual chest-throwing that I answered in the cipher-code of the higher priesthood and heard Martin's self-abasing and worshipful 'Oh.' As for Charmian, I felt that in a new way I had proved my right to her; and I was aware of another feeling, namely, that she was a most fortunate woman to have a man like me . . ."

—Jack London, from
The Cruise of the Snark

APPENDIX
ABBREVIATIONS

Appendix: Abbreviations

a_0, a_1, a_2	Polaris sight corrections
Add'l Corrn	Additional correction
Alt. Corrn	Altitude correction
A.M.	Ante meridian (before noon)
AP	Assumed position
App. Alt.	Apparent Altitude
Co-Alt.	Coaltitude
Corr Hs	Corrected sextant reading
Corrn	Correction

d	Altitude difference
d corrn	Correction for change in declination
Dec	Declination
DR	Dead reckoning
DST	Daylight saving time
E	East
GHA	Greenwich hour angle
GMT	Greenwich mean time
GP	Geographic position
Ha	Apparent altitude
Hc	Computed altitude
Ho	Observed altitude
H.P.	Horizontal parallax
Hs	Sextant altitude
IC	Index correction
Int	Intercept
L	Lower (limb)
LHA	Local hour angle
LOP	Line of position

MHz	Megahertz
N	North
N.A.	*The Nautical Almanac*
P.M.	Post meridian (afternoon)
S	South
s	second
SHA	Sidereal hour angle
U	Upper (limb)
v corrn	Correction for irregular orbital motion
W	West
WE	Watch error
WT	Watch time
Z	Azimuth (180° east or west)
zd	Zenith distance
ZD	Zone description
Zn	True azimuth (360°)
ZT	Zone time

CAPTAIN JOSEPH E. THOMPSON

A Mid-Westerner by birth, Captain Joe Thompson gravitated to the sea in the late 1950s. A graduate of the University of Florida, he has served in the United States Air Force, earned a master's degree from the American Graduate School of International Management, and worked in banking in New York, Brazil, and Florida. The holder of a Coast Guard Ocean Operator's license, Captain Thompson enjoys offshore cruising and has long felt that a jargon-free, step-by-step manual has been needed by students of navigation. CELESTIAL NAVIGATION, Captain Joe Thompson's Cookbook Method is that book.